Joan of Arc

Joan of Arc

A STUDY IN CHARISMATIC
WOMEN'S LEADERSHIP

ANTON KAISER

black hills books press

Front cover: "Entrance of Jeanne d'Arc at Orleans" by Jean-Jacques Scherrer.

Back cover: Fleas brought the bubonic plague which infected half the European population while the University of Paris declared Joan's similar "poisonous influence has spread so far that nearly the whole of the Christian flock of the west has been infected." Bees and bee-keepers were often associated with heretics in medieval folklore, and butterflies symbolized winged souls; spiritual afterlife resided in these flutterings. Thus, to float like a butterfly and sting like a bee implies a resurrected spirit with the sting of a heretic. These were the insects of the hour.

FIRST U.S. PRINT EDITION

Hardcover ISBN-13: 978-1-945333-08-8
Softcover ISBN-13: 978-1-945333-09-5
Library of Congress Control Number: 2017953120

Published by Black Hills Books Press, Rapid City, South Dakota

Available from Amazon.com and other book stores
Available from the author
Email joanofarc@blackhillsbooks.com

Printed in the United States

ACKNOWLEDGEMENTS

After decades of researching Joan of Arc there are plenty of people to thank, but most of all I want to thank America's library system. Our librarians are a national treasure. Even before the emergence of the internet these remarkable professionals could find anything anywhere and deliver it with remarkable efficiency. I hold the librarians of America in deepest respect and profess my most heartfelt thanks. Also during this study, I was privileged to observe first-hand the capabilities and potentials of our women in the U. S. Army. Not all of them were capable of serving in the infantry, but many could and did meet the physical and mental leadership standards required by this tough profession. Their example allowed me to keep an open mind and believe in the same possibilities in Joan of Arc. Others, like the members of the Black Hills Writers Group, offered useful critiques during brief readings of selected portions, testing my ideas against a published audience. Finally, I called on an expert team of editors from Black Hills Books Press who professionally scrutinized every word, dot, and dash, bringing my work to a polished state of completion. Dr. Dan Syljuberget, Dr. Chris Bull, Mimi Lancsak, and Bonnie Norrod were especially insightful, patient, and knowledgeable in reviewing and commenting on my work. The leader of this generous and talented team, the indomitable Barb Evenson, deserves special thanks for guiding this project, and especially for donating as much time, effort, oversight, and expertise as she did, given the otherwise enormous demands on her talents. Again my thanks to all.

Contents

PREFACE

How did she do it!? This simple question led to over thirty years of research and study about Joan of Arc.[1] As a professional army officer, I found the hundreds of books written about her unconvincing. My criterion for this judgement was simple: Would I have followed this seventeen-year-old girl into combat? Consistently my answer was no. But her history is undeniable and men did follow her into battle. Why? There had to be something beyond religious fervor driving these soldiers to risk their lives at her command. From my own experience, most soldiers, even medieval soldiers, would not fight and die simply because a religious female waved a banner and claimed it was God's will. Soldiers are universal and often more cynical than credulous. Thus I set out to find the answer, to figure out how it all happened, and, indeed, to convince myself, if possible, that I too would have followed her leadership.

My search ended decades later, but only after I pieced together, year-by-year, acceptable answers to this puzzling mystery; answers that gave Joan legitimacy and led to the conviction that I too would have followed her into combat. I then realized others might find these answers useful and set about developing the history of Joan's leadership as I understand it. In many ways it is different than any other book written about Joan, maybe because it is only the second book written about her by a professional *American* army officer, and the only one written by a professional American *infantryman.*

Few people have ever confronted the assignment that brought Joan of Arc to my attention. Back in the early eighties, one bright and early morning, I found myself crouched defensively behind a highly shined Pentagon conference table. To my right sat Major General Maxwell Thurman, Deputy Chief of Staff for Personnel (DCSPER), and across from him sat Brigadier General Norman Schwarzkopf, his deputy. General Thurman

[1] Why thirty years? Because I have written three complete biographies of Joan. The first two efforts I threw out, because they were unconvincing. The third I hope to publish.

later commanded Operation Just Cause, the Panama invasion, which strike force included the Seventh Infantry Division (Light) from Fort Ord, California, where I labored as the Inspector General at the time. General Schwarzkopf later commanded operations Desert Shield and Desert Storm during the first Persian Gulf war. Behind them, however, sat an even more intimidating group of warriors, political females with a look of blood in their eyes, known as WITA, the Women In The Army study group. As best I knew, WITA was an offshoot of the most feared organization known to the military at that time: DACOWITS, the Department of the Army Committee On Women In The Service.

A smattering of other military onlookers sat at the far end of the mahogany table, while those less secure bunkered themselves in the smaller recesses cluttered about the room, all in a futile effort to avoid the WITA weaponeers. Maybe two dozen people were in attendance. I was scheduled to brief a highly controversial subject: whether women officers should be included as members in the combat arms branches of the U.S. Army.

The American army divides itself into three categories of service: combat, combat support, and combat service support branches, which at the time were defined by their normal operating distance from the frontlines. But the new airland battle doctrine then emerging blurred these boundaries. The combat support and service support branches, traditionally more distant from the dangers of direct combat, were rapidly becoming, by technological fiat, increasingly programmed to operate inside high-risk, close-in, tactical environments. Meanwhile, in 1976, Congress opened the army doors to include more women, traditionally assigned the much safer support and service support branches. As a result, the United States Military Academy, West Point, released its first class of female graduates in 1980, and college ROTC's followed suit. By the early eighties, army numbers were beginning to bulge with these new female recruits who expected equal treatment. Where would the army place all these women officers? It fell to me, a young infantry Major, to study and then to recommend the army's future policy concerning this highly volatile and contentious issue.

Why did I get this assignment? I suspect because no one in the army wanted it, including me, but I had been ambushed. Without consultation or warning, I was suddenly pulled from the field where I was exercising with the Fifth Infantry Division (Mechanized) at Fort Irwin, California, at the new National Training Center, and ordered to report to Washington, D.C., within thirty days. When I finally reached a telephone to call my infantry branch, thinking I could argue my way out of this mistreatment, I was bluntly told it was an emergency, and besides it was my time in the barrel, meaning we all had to serve in the capital at some time. Of course I pulled all the strings I knew trying to escape this assignment, which proved to be way too few strings, and which inadequacies proved to be my true qualifications for the job.

Soon I was flying all over the United States, meeting with various groups—colleges, training bases, and West Point—in an attempt to solve a problem nobody wanted to discuss, and which solution nobody wanted to contemplate. As one might expect, I was less than warmly welcomed and often poorly accommodated wherever I went. The country was awash in a full-fledged and highly charged feminist movement which by its very nature was hunting and consuming animals like us; that is, army officers who still saw their profession as an exclusive all-male club, a masculine oasis sheltered from the otherwise troubled sands and constant whirlwinds of female aggressions. Nevertheless, in my spare time I read the government authored Athena Study, which recommended the admission of women into West Point in 1976, as well as the histories of the Amazons, Greek mythological female warriors, and more modern military histories, especially those of the Soviet and Israeli armies, all in search for a silver bullet: the "right" answer to whether women should serve in our modern combat arms branches.

I had prepared a trick presentation, which style is not wholly approved of in the staid and imperial world of Pentagon techniques; nobody likes to be fooled. Furthermore, political correctness existed even then and I knew only too well the position I was expected to take. So for ninety percent of my briefing, using two dozen vu-graphs, I talked the politically correct army talk and walked the politically correct army walk. Besides, I was a chauvinist, as my mother so often reminded, and represented well the infantry culture of my longstanding male heritage. Meanwhile, other male officers listened smugly (secure in the back recesses of the room), wondering how I was going to get myself off the hook, even as WITA papers shuffled viciously in the background making ready their attack against my anticipated conclusion. But then, and by using the very logic every military officer in the room had been nodding to support, I presented my last two slides which recommended, unequivocally, women assignments in the combat arms branches of the American army.

Besides the more volatile issues of career development, or professional opportunities for female officers, or debates about physical strength requirements, or the more emotional issues of potential female body bags in the event of war, army numbers had been shrinking since the Vietnam era, even as these new female officers flooded the ranks. Their numbers had reached 12% of the army force structure (today it is 16%) while continuing to grow. With this basic information we used computers to age the army, projecting the trends. As it turned out, women were already consuming nearly 50% of all officer positions then available within the much smaller combat support and combat service support branches and with any further increase in women officers or decrease in force structure these leaders would quickly become the dominant sex in these branches, while combat arms branches remained 100% men, dividing our army leadership by sex. So what was wrong with that, one might ask?

The problem it seemed to me was more than just force structure and demographics. Branches within the army have their own rivalries. Men in the combat arms often refer to members in the combat support and combat service support branches, who function more commonly in the much safer rear areas, as "clerks and jerks." These are the soldiers who escape the cold wet foxholes, who eat three meals a day, bath in hot water, avoid artillery bombardments, and rarely hear a bullet fired in anger. Then they ask for everything in triplicate when a tired, sore, wet, hungry, cold, and scared combat soldier requests a new pair of boots worn out from his endless days and months of fighting. By example, during World War II (Rick Atkinson, *The Day of Battle*, 452), the rear echelon American service troops were "the most hated folk in Italy, the Germans running a poor second." Likely this animosity occurs because the combat arms branches suffer almost 90% of the casualties, as they did during WW II. Further, this animosity can reach as low as battalion level within the combat arms units themselves (Charles B. MacDonald, *Company Commander*, 264).

Thus, the army in the early eighties was placing itself in the very politically sensitive position of stereotyping women officers as "clerks and jerks," subject to all the expected animosities usually inherent in the attitudes of frontline soldiers toward their rear echelon support, especially during times of war. We were willing to let these women cook our food, wash our clothes, buy our groceries, fix our radios, type our papers, haul our ammunition, and fuel our vehicles, but not fight our wars. In domestic terms, women could be our housekeepers but not our breadwinners. We either had to open the doors to the combat arms branches or create a starkly divided and potentially explosive prejudicial system. Of course my analysis entered into the forbidden realm of social politics as concerned the military and ignored the even more crucial issue of combat readiness (although dividing the army by sex seemed to me to jeopardize the morale and teamwork so essential to an army), but, as I suspected, no one was ready or willing to challenge the potential combat capabilities of women. This approach to the problem (using a football analogy) was somewhat like a hand off to the halfback, with a reverse to the tight end, a pitch back to the quarterback, followed by a "Hail Mary" pass to the wide receiver—hoping nobody was looking for it. I ended the briefing and waited to see if I had completed the pass or would be run out of the stadium for calling a foolish play.

To my surprise, General Thurman responded with great delight and could hardly contain himself. He turned on the stunned room and enthusiastically asked for questions or comments. No one knew what to say; they had all been caught off guard. "It's done, then," he said, and quickly left the room. It happened so fast no one could imagine the implications. A few weeks later the Judge Advocate General (lawyers) blessed the final policy paper and General Thurman signed a simple one page document making it official: women officers would serve in the combat branches of the Engineers, Aviation, Field Artillery, and Air Defense Artillery, but for now, not in the Infantry or Armor—in essence, we had split the baby in half.

Internally, it was not a popular decision and set in motion a year long political struggle inside the army designed to undo the unthinkable. A year later I was called to brief the new DCSPER, who had replaced General Thurman, and I essentially gave the same presentation. This time however the debate was intense and without decision. Days later I was abruptly advised that women officers would be taken out of the combat arms branches; it was judged a bold policy doomed to failure. Almost immediately I received a bevy of "I told you so" telephone calls from my equally chauvinistic but good natured Pentagon buddies. Still, I believed I was right, even if I was being called the infantry's first feminist.

A week went by while we all tried to adjust to another policy change. Numbers had to be recalculated, career programs revised, and toilets inventoried. Then, out-of-the-blue, the phone rang again, this time with a very distraught voice at the other end. The new DCSPER had sent a memo to the new Vice-Chief of Staff of the Army advising him of his decision to take women officers out of the combat arms branches, but the new Vice-Chief was none other than that same but recently promoted Lieutenant General Thurman! Of higher rank and position, he instantly reversed the new DCSPER's decision.

These were historic events and to this day female officers serve in the combat arms branches of the American army, but it was a near thing. General Thurman, who later captured Panama and blasted rock music at the drug dealing Manuel Noriega to pry him out of hiding, can be given all the credit. He has since passed away.

I remember hanging up the phone that day and jokingly wondering if the remarkable Joan of Arc had had anything to do with what transpired; after all, I had invoked her spirit and she was known to achieve the impossible. In fact, I began to wonder just how it was that chauvinistic men had fought and died for her, men like me, infantrymen who could hardly imagine the possibility of fighting to the death at the command of a seventeen-year-old woman. But then, I realized, I had been confusing Joan of Arc, a real person, with her fabled myth and legend. Unlike Athena, Joan had been real, despite the almost surrealistic accounts historians make of her life. I resolved someday to rectify this disparity between fact and legend; after all, why shouldn't women join us in our foxholes and tanks: maybe we had gotten it only half right. When I finally retired I set about to re-study this issue and there was only one place to begin: who was the real Joan of Arc and why would chauvinistic medieval men agree to fight under her command?

Almost thirty years later, in 2012, Brigadier General Laura Richardson became the first female officer in the American army to be appointed as an assistant division commander in a combat arms unit, the 1st Cavalry Division. She had risen to that position through her combat arms branch—Aviation. Additionally, in 2016, the Army opened the Armor and Infantry branches, all of its positions, to women. Enlisted women for the first time have the opportunity to serve as cavalry scouts, armor crewmen, infantrymen, and indirect-fire artillerymen (or will we add the terms infantrywoman and artillerywoman to the army's lexicon?). As a result of this change, about 90,640 positions were opened to women in the Army. Within these same branches, the Army will also offer junior officers and junior NCOs the opportunity to transfer branches or reclassify into these occupations as a way to build a cadre of experienced female soldiers prior to the arrival of soldiers new to the Army.[2]

[2] http://www.army.mil/article/105814/Army_describes_plans_for_integrating_women_into_combat/.

The female cadets graduating from West Point in 2017 selected the following combat arms branch assignments for their officer commissions: Engineering—24; Field Artillery—14; Aviation—12; Air Defense Artillery—6; Armor—4; Infantry—4; for a total of 64. Ninety entered the combat support and combat service support branches.

Also in 2017, the Marine Corps is expected to assign its first female infantry officer to a combat unit.[3]

Although my rationale for recommending female entry into the combat arms branches centered on Army teamwork to win wars, leadership was a strong second bulwark behind my reasoning. During my study, Joan of Arc stood out as a combat leader and can be given substantial credit for the support I retained as a backup to my recommendation. Nevertheless, much of her history still remained a mystery from a professional military point of view. This book solves that mystery.

After finally deciding on publication, I also decided to avoid the typical biography and concentrate only on her leadership and how she acquired her remarkable military talents, talents often deemed miraculous but which actually derive from traditional military leadership doctrine as established throughout history. A full biography of Joan will follow with unique revelations also not found in any other book about Joan of Arc.

Anton Kaiser

[3]https://www.washingtonpost.com/news/checkpoint/wp/2017/09/21/for-the-first-time-the-marine-corps-expects-to-have-a-female-infantry-officer-among-its-ranks/?utm_term=.9119af4ca298

Reading histories of aristocratic times, and especially those of antiquity, it seems that man has only to learn self-discipline in order to become master of his own destiny and to rule his fellows. In perusing the histories written nowadays [1830s], one would suppose that man can have no power over himself nor over his surroundings. Historians of antiquity taught the rules of command; those writing today give only a pale lesson on how to obey.

Alexis de Tocqueville
Democracy in America

CHRONOLOGY

1337 The Hundred Years War begins (April).

1347 Bubonic plague arrives in Europe (October).

1412 Joan of Arc is born in Domremy, France (January).

1415 The Battle of Agincourt (October).

1419 Assassination of John the Fearless (September).

1420 The Treaty of Troyes is signed (May).

1422 Henry V dies (August).

Charles VI dies (October).

1428 The blockade of Orleans begins (October).

1429 Baudricourt sends Joan to Charles VII (February).

Joan arrives in Chinon (March).

Joan questioned at Poitiers (March—April).

Joan arrives in Orleans (April).

St Loup, St. Le Blanc, the Augustins, and the Tourelles are taken, and the English retreat (May).

Jargeau, Meung, Beaugency are taken and the English are defeated at the Battle of Patay (June).

Charles VII is crowned king at Reims (July).

Attack on Paris fails (September).

Attack on St Pierre le Moutier succeeds (October).

Attack on La Charite fails (November).

Joan and family ennobled by Charles VII (December).

1430 Joan is captured at Compiegne (May).

Joan is taken to Rouen (December).

1431 Joan's trial for heresy begins (January).

Joan is burned at the stake (May).

1453 The Battle of Castillon ends the Hundred Years War (July).

1456 The nullification trial declares Joan's innocence (July).

1920 Joan is canonized by the Catholic Church (July).

Introduction
Framing Her History

> *Because half a dozen grasshoppers under a fern make the field ring with their importunate chink, whilst thousands of great cattle, reposed beneath the shadow of the British oak, chew the cud and are silent, pray do not imagine that those who make the noise are the only inhabitants of the field: that, of course, they are many in number, or that, after all, they are other than the little, shriveled, meager, hopping, though loud and troublesome Insects of the hour.*
>
> Edmund Burke, *Reflections on the Revolution in France*

Biographers often waste countless pages and endless chapters describing the turbulent events and chaotic characters preceding Joan of Arc. By example, Rupert Evans in a book of thirteen chapters devotes nine of them to explaining the European history immediately preceding her emergence as a revolutionary force in history.[1] In a like manner, Andrew Haggard devotes eighteen of his twenty-six chapters to prior events.[2] The following introduction severely compacts this early French chronology to save space for Joan of Arc and her soldiers.

Will Durant eloquently and succinctly described the medieval age as follows:

---

[1]Rupert Evans, iii-iv, answers, "...if I had not first clearly explained the state of France, and the cause which reduced that country to the dreadful condition it was in...they could not have perfectly understood the amazing value of the Maid's services." By fact, Joan's name is often attached to book titles that have very little to do with her actual history—for example, Haggard's *The France of Joan of Arc* or Wood's *Joan of Arc and Richard III*—but the name sells.

[2]Haggard's chapters on Joan are woefully inaccurate and more poorly researched than his interests in the history prior to Joan.

> So "virtus," manliness, was restored to its Roman masculine sense after a thousand years of Christian emphasis on feminine virtues. Chivalry, despite its religious aura, represented a victory of German, pagan, and Arab conceptions over Christianity; a Europe attacked on every side needed the martial virtues again.....These rough knights drove the Moors into Grenada, the Slavs from the Oder, the Magyars from Italy and Germany; they tamed the Norse into Normans, and brought French civilization into England on the points of their swords. They were what they had to be.[3]

Into this martial world on Twelfth Night,[4] January 6, 1412, a little girl was born who would not only embody this medieval masculine spirit but also its feminine soul of Christian origin. Joan of Arc synthesized history and fused its martial ardor and religious vulnerability deep into the hearts of humanity. According to Joseph Delteil, "She was the marriage of Earth and Heaven."[5] She was what she had to be.

Each century brings its own trials, and the origins for Joan's persecution really began as early as 1336 when the count of Flanders, based on paranoid suspicions of economic subterfuge, jailed all English merchants then residing in his Flemish lowlands. Within weeks the English king, Edward III, responded in-kind by arresting all Flemish citizens living on his island. As so often the case, however, Edward further retaliated by banning the export of English wool, vital to Flanders' textile trade. These mutual incarcerations and economic retaliations were the origins for war.

The count of Flanders, unable to withstand Edward's challenge to his textile industry, quickly buckled to its pressures, blaming France for the whole affair. The French king, he claimed, had started the troubles between England and Flanders by surreptitiously undermining the commercial and financial interdependence flaming so richly between these two

---

[3]Durant, *The Age of Faith*, 575.

[4]January 6 is the Feast of the Epiphany, or Twelfth Night, which celebrates the Magi's rendezvous with Christ in a Bethlehem manger and is commonly reported as the date for Joan's birth. Likely, this date is myth, as is so much of Joan's history, because no one really knows nor did Joan herself know the exact year, month or day she was born. History, a world unto itself, has wrapped her life in a labyrinth of fabrication and mystification, even while her true life is miracle enough. (Interestingly, in Shakespeare's *Twelfth Night*, the alleged night of Joan's birth, the main character, Viola, also disguises herself as a male named Cesario, and like Joan also serves as a messenger, but brings "no overture of war." Given Shakespeare's talent for subtlety, one must wonder if he wasn't toying briefly with his Joan of Arc from *Henry the Sixth*.)

[5]Delteil, 148.

economic partners. In fact, southwestern France *had* been interfering with the wine and wool trade between England and Flanders.[6] England therefore declared war on France, a war not caused by religion, for which so many wars are blamed,[7] but by politics and its faithful partner, economics. Unexpectedly this quarrel lasted one hundred and sixteen years and came to be known as the Hundred Years War, which conflict consumed the life of Joan of Arc.

A year later, in 1337, while Edward busied himself organizing political alliances with the Germanic states and Flanders, King Philip of France struck back by annexing the English ruled continental province of Aquitaine. In further escalation, Edward declared himself the rightful king of *all* France,[8] eventually setting sail to conquer his new estates. In 1340 he destroyed the French fleet at the Battle of Sluys and the war was official.

Joan inherited this divided world. On one side of the European conflict stood England, aligned with Ireland, Flanders, Portugal, Germany, Italy (less Naples and Savoy), Poland, Hungary, and Bohemia, while on the other side stood France, Scotland, Spain, Cyprus, Naples, and Savoy.

Also during this century the powerful Catholic Church divided itself, suffering another Great Schism. Pope Clement VII resided in Avignon, France, while Pope Urban VI fought back from Rome, Italy, both claiming the seat of Peter and each excommunicating the adherents of the other. Everyone stood heretic.

After further decades of fighting, Joan's own century dawned in 1400 with the death of Geoffrey Chaucer, famous for his *Canterbury Tales*, an appropriate precursor to Joan's own adventures. Also occurred the ordainment of John Hus, a Bohemian priest whose alleged heresy would eventually reach deep into the continent of Europe, preconditioning Joan's own religious unorthodoxy. Also occurring was the assassination of Richard II, King of England, whose premature death fatefully framed Joan's historical presage. Finally there arrived the birth of Johann Gutenberg whose printing press would end Joan's historical era. But the most important event during this year occurred when Yolande of Aragon married Louis II of Anjou. Yolande would reconstruct the French monarchy using Joan as one

[6] Ashley, 503.

[7] "When states go to war, it is usually for a variety of generally secular causes. Religion may be quickly involved, but only because it is such a potent instrument for energizing society and motivating troops." Wade, 241.

[8] Authors often begin the Hundred Years War with Edward's claim to the French throne, which is a simplification.

of her primary architects, which claim is one of the more controversial theories in Joan's history.

During the next twelve years Europe continued its convulsions in preparation for Joan's birth. The religious turmoil resulting from the Great Schism diluted the faith and confused the faithful. Adding to this chaos, the immensely influential universities of London and Paris, along with their supporting English and French monarchies, split their loyalties between the two competing Popes; England sided with Rome and of course France sided with Avignon. As Friedrich Heer so astutely observed, "The two most powerful estates, the nobility and the clergy, cut themselves off from the masses and became closed societies; so did the intelligentsia, jealous of their university learning and privileges."[9]

The various medieval guilds and their merchant peddlers, an emerging middle class, added to this turmoil by joining in this economic chaos and by challenging both the monarchies in division and the universities in dissension. Not to be outdone, the lowly peasants brought occasional bloody revolts against all three state divisions, further stratifying the social classes. Joan would begin her adventures at the bottom of this class structure and end at the top of it.

Although horses and wagons were the primary means of medieval transportation, the Atlantic states of Spain, Portugal, and England, along with the Mediterranean states of Italy, Greece, and France, together with the Hanseatic League governing the northern sea lanes of Europe, all began expanding their shipping fleets and stretching their trade routes. Merchant fairs became increasingly common, introducing new spices, ideas, inventions, and epidemics. As these trade routes grew, so too did the need for political compromise, meaningless treaties, and profuse entreaties, until the markets strangled themselves, often leaving tattered food baskets hanging hopelessly on the bony arms of the peasants. Although a perilous seasonal agriculture remained the primary means of subsistence, a burgeoning wool and textile industry added sheep, herding, wool, and the art of spinning, weaving, and sewing to the meager labors of life, of which Joan would partake and become expert, especially at spinning and sewing.

Consequently, by the early fifteenth century, Europe had become a land of plenty; that is, plenty of forests, swamps, wolves, and poverty. The rich got richer and the poor poorer; governments taxed and abused their

[9]Heer, 24.

citizens; merchants kept their thumbs on the scales; and churches reached out to rob the common man of his last penny even while he knelt to ask forgiveness for his sins. As the anonymous bourgeois author of the medieval *Journal of Paris* so aptly put it, "the great all hated each other, the middle class were burdened with taxation, [and] the very poor could not earn a living." In short, the world was much as it has always been. Of course the military profession, which eventually embraced Joan, found a cozy home amidst all these frustrations, as it does to this day.

Before the turn of the century, John Wycliffe had preached an independent English church attempting to advantage the Catholic schism by translating the Bible from Latin into English, just as Martin Luther would later translate it into German further igniting the Reformation. However, in 1403, the University of Prague quickly condemned Wycliffe's writings. Notwithstanding, in 1407, Jerome of Prague returned to that same university, having studied Wycliffe in England, and joined with John Hus to incite the Bohemian academics against the Catholic canons. Meanwhile, Pope Urban VI died. Thus when the Colonna family sacked the Vatican in 1406, it was Innocent VII who suffered the heart attack, and Gregory XII who was elected the new Pope in Rome. Wycliffe's ideas would find their way to Joan of Arc.

But these ordeals were only precursors to the most significant event, which also occurred in 1407, that would truly affect Joan's life. In that year John the Fearless, duke of Burgundy, assassinated Louis, duke of Orleans, using a gang of thugs to catch his victim in a Paris back alley where they cut off one of his hands and crushed his skull. Louis had been governing France on behalf of his brother, Charles VI, whose sufferings from an intermittent madness had begun in 1396. Louis' death gave rise to a loyalist faction called "Armagnacs," royalists whose protests against this brutal assassination ignited a French civil war. Accordingly, the butcher's guild supported the duke and his Burgundians while the carpenter's guild supported the Armagnacs. Thus France joined much of the European continent, internally divided and warring against itself, as were also Spain, Italy, and Bohemia. The task of "nation building" would fall to Joan of Arc.

Also in 1407, the Council of Pisa met to decide the true Pope, but, as usually happens with rule by committee, it succeeded only in complicating the issue by electing a third Pontiff, Alexander V. The next year Alexander V died and John XXIII became Pope. Meanwhile in 1410, John Hus, who had become rector of Prague University, protested the Catholic Church's burning of Wycliffe's books which resulted in his excommunication for heresy. Upon this scene, in 1412, Joan entered the world.

A year later, Simon Caboche, a bourgeois butcher, joined with the University of Paris in another rebellion against the French crown. Despite the resulting Cabochien Ordinance, increasing the political powers of the Estates General, the movement collapsed that same year. The citizens of Paris soon realized that an even more brutal regime, as represented by a combination of butchers and academics, could emerge accidentally when trying to replace an already existing brutal regime with a more intellectual alternative, regardless of how appropriate the political partnership may have seemed (which reminds us of our modern intellectuals and their disparagement of the Shah of Iran resulting in his more treacherous replacement, the Ayatollah of Iran, whose regime established the model for the state sponsored terrorism now plaguing the world—another hundred years war?).

As so often happens in history, the year 1415 joined several momentous events. Henry V, who had been crowned king of England just two years prior, promptly invaded France, winning the Battle of Agincourt and killing off a large number of French nobility while also capturing Charles, duke of Orleans, whose imprisonment would play a prominent role in Joan's life. Also in that year, the Council's ideas of reform burned John Hus at the stake for heresy, which fate would also find Joan of Arc. Meanwhile, the Council of Constance, which convened to reform the Catholic Church, deposed John XXIII and Benedict XIII (Benedict XV would eventually canonize Joan in 1920), and forced Gregory VII to resign—the last Pope to resign until the resignation of Benedict XVI in 2013. Thus, the world was without a Pope until 1417 when the Council elected Martin V to end the Great Schism, which had begun in 1378. The death of Martin V would complicate Joan's problems during her heresy trial fourteen years later. Meanwhile, in 1416, Jerome of Prague was also burned at the stake.

All of these events signaled a revival of the Church Inquisition,[10] while the death of Hus ignited a Bohemian civil war led by the Hussites who spawned five Church crusades against them during Joan's short lifetime.

In 1418, the now rebel duke of Burgundy, John the Fearless, attacked and captured Paris, slaughtering 2,000 defenseless Armagnacs inside the

[10] According to Barstow, 102n, from 1428 to 1438 approximately two hundred and sixty people were brought to trial for some form of heresy, many of them put to death. Goldstone, 174-75, claims there were seventy witch trials between 1410 and 1430.

city. Given the severity of our modern response to the loss of life from terrorist attacks, the Armagnac reaction should be no surprise. Thus there followed a year later the greatest event that would truly control Joan's destiny. The sixteen-year-old boy who would become king of France, Charles VII, retaliated against the Paris massacre by assassinating John the Fearless (not unlike our assassinations of Yamamoto in WW II in retaliation for the 2,400 deaths at Pearl Harbor or our assassination of Osama bin Laden in retaliation for the 3,000 deaths surrounding the twin towers). Charles had managed to arrange a meeting under the guise of ending the internal rift between the royalist Armagnacs and rebel Burgundians and took advantage of this opportunity to ambush John the Fearless on the bridge at Montereau. John's death however gave rise to his son, Philip, who openly allied his Burgundian party with England in another pledge of retaliation. These continental French rebels together with the English invaders would become Joan's avowed enemies.

In the meantime, the religious orders, those in nearest contact with the common populace, also divided. The Dominicans sided with Philip and his Burgundians, while the Franciscans pledged their loyalty to Charles VII and his Armagnacs. Everywhere, it seemed, the world was in factions.

In 1420, when Joan was about eight-years-old, Henry V parlayed his crushing victory at the Battle of Agincourt and his brutally successful siege of the French city of Rouen into the Treaty of Troyes, a devastating document supported by the French queen on behalf of the mad Charles VI, awarding all of France to England beginning with Henry V's progeny. Charles VII's sister, Catherine, married Henry V in accordance with this treaty and the birth of Henry VI a year later joined the English and French monarchies, sealing the death of France. Henry VI was destined to become the first monarch to wear both crowns. Without Joan, Gallic destiny lay firmly in the claws and jaws of the Norman/Anglo-Saxon lion. Any political future for an independent France seemed legally hopeless. By year's end, the English king, Henry V, entered and occupied Paris.

Nevertheless claimants for Charles VII challenged this treaty, and the continental fighting continued resulting in wins and losses on both sides, yet seeming to favor the allied English and their Burgundians. Finally, in 1428, two significant events occurred: the English dug up Wycliffe's bones and threw them into a river, which signaled the same fate for Joan's own charred ashes upon her death; and the English launched a major military offensive allegedly designed to drive the last vestiges of French royalty from off the continent or to beyond the Pyrenees mountains and into Spain. This English attack struck both to the west of Paris and south to Orleans where these foreign invaders, together with their Burgundian allies, established an alleged siege of that ancient city. Subsequent events around the city of Orleans would make the seventeen-year-old Joan of Arc an historical legend.

Against this backdrop of countries at quarrel, institutions in conflict, kings in contest, Christians on crusade, merchants in conspiracy, and peasants in revolt, little Joan of Arc grew up quietly in the fields and pastures outside Domremy, France, a small village resting quietly on the eastern frontier of the royal realm.

One day at play, age twelve or thirteen, Joan ran so swiftly across the fields her astonished friends cried they could not see her feet touch the ground. At the end of this race, a stranger intercepted her with a message—her mother wanted her at home. But when Joan returned to her small village, she was scolded for leaving the field and her chores—no one had sent for her. On this occasion and at this moment legend changed Joan's life.[11] Suddenly, while standing in the family garden, she experienced a bright light and from within this brilliance appeared another stranger, a messenger who was not only a vision but who also brought a vision—a vision that would kill Joan but save France. This messenger, according to Joan's own testimony, came from God in the voice and form of the Archangel Saint Michael who arranged for the voices and visions of Saint Catherine and Saint Margaret to guide Joan toward her dramatic and traumatic fate.

The history of Joan of Arc is one of exhilaration and heartbreak. In just three short months in 1429 she would rise from the quiet pastures of peasant obscurity to thunderous acclaim and military fame, electrifying the European continent. Her quick series of military victories—a lightning bolt across the battlefields of France—would astound both enemies and allies alike, leading to a new nation. Joan and her soldiers were Edmund Burke's insects of the hour and their brief noise in the pastures of France would change history. But after crowning Charles VII as the true king of France, Joan's legacy would slide into an oblivion of loss, capture, and trial, leading to an early death at age nineteen. Hers is a history without precedent, from a personality without compare; a history that reveals more about ourselves as members of the human species than we would normally like to know, yet

[11]This story is also likely a myth in a letter by Perceval de Boulainvilliers in Lowell, 30; Lang, 40-1; and Harrison, *Joan of Arc*, 23.)

lessons worth learning; and maybe that is why she existed—as a starburst of optimistic inspiration for the human spirit.

Chapter One

SETTING THE STAGE

Before raising the curtain on this three-act melodrama, we need to set the stage for Joan of Arc's amazing life. Although not alone in this tragedy, she is soon imagined the solo performer as we delve deeper into her history. As with all of us, she lived inside a cultural environment, a medieval setting of social customs, conventions, traditions, and morals, which provide the backdrop for her myriad leadership talents, a very powerful set of martial instincts encompassed within a sophisticated collection of influential social abilities all skillfully applied to the conditions of her time. These instincts and talents are of classic origins having survived the scrutinies of history, especially in the military art, and enjoy universal application.

As currently acknowledged in U.S. Army doctrine, "Leaders operate on instinct that has evolved from what they have seen. What leaders see others do sets the stage for what they may do. Modeling these attributes of character defines the leaders to the people with whom they interact."[1] In other words, our personal instincts, as well as modeling or duplicating the leadership instincts of others, create our individual leadership traits. Joan's extraordinary leadership talents had

> **"Leaders operate on instinct that has evolved from what they have seen. What leaders see others do sets the stage for what they may do. Modeling these attributes of character defines the leaders to the people with whom they interact."**

[1] Army Leadership, 6-10.

to come from somewhere, but where? In this study we will search them out, something unfortunately not found in any other books about Joan of Arc.

Jules Michelet's classic, *Joan of Arc*, written in 1841, is said to be the most influential book written on Joan in the modern era,[2] and maybe so, since most of his intellectual judgments remain with us today, held hostage within a multitude of Joan of Arc biographies despite frequent factual errors and exaggerations; thus, Michelet's work represents the starting point for a biographical sprint through an intricate epoch that includes a bevy of nonsensical myths. By combining Michelet's heavy historical influence with the modern definition of leadership, our search for Joan of Arc's executive abilities begins.

FAITH

REASON

EMOTION

INTELLECT

In the first paragraph of his introduction to Michelet, written in 1926, Gustave Rudler makes the case for embracing faith over reason and emotion over intellect when reading Michelet's history of Joan, offering two sets of sensations and values often directly opposed to each other. The battle between religious faith and academic reason began as early as Aristophanes in ancient Greece, continued with Peter Abelard in the twelfth century,[3] intensified in the debates between Luther and Zwingli during the Reformation, reached its zenith during the Italian Renaissance, and has been ongoing ever since; meanwhile, the battle between our volatile emotions and cold intellect probably began in the Garden of Eden sometime before man's—or woman's—first campfire. But soon discovered, we find that Rudler was on to something: faith, reason, emotion, and intellect are what define the mystery if not the history of Joan of Arc.

Emerging out of the eastern edge of France near the Vosges Mountains in the tiny village of Domremy, the medieval writings about this remarkable girl reveal a story of faith, an unreasonable faith to some, but a faith that changed European history. This faith however is not so much religious faith in the traditional sense, but a devotion or a duty to that religious faith, best described as a faith in her faith. Rudler identifies this faith as "fideism" (fee-day-ism), which the modern dictionary defines as a belief in a religion

---

[2] See Richey, 9, quoting from Margolis, *Joan of Arc in History*, 128-29.

[3] "Faith, so easily confused by emotional excitement, should not be allowed to stand permanently in the path of reason; and reason for its part should not use the intellect as the means of solving the final mysteries of faith, which were inaccessible to reason." Heer, 112-13.

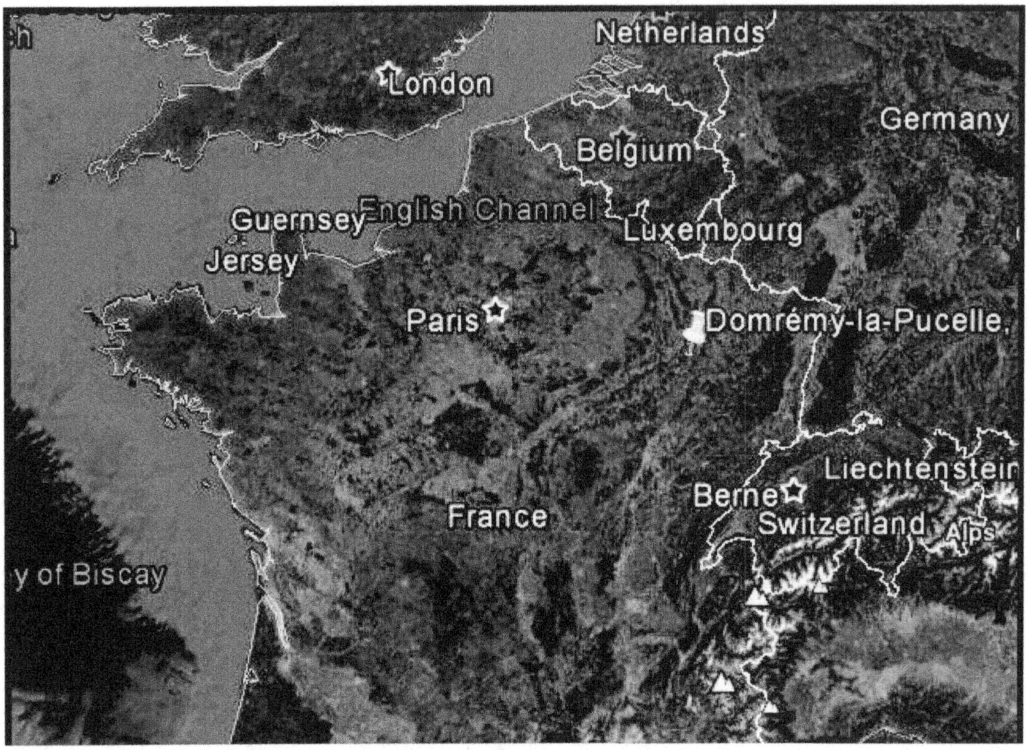

Domremy, France. (Image@2016 Google Earth)

or philosophy that transcends rational explanation and is accepted on faith alone. In other words, fideism is a commitment to faith, a faith in the duty to one's beliefs or to one's religious faith. More than religious faith alone, this is the faith we mean when referring to Joan of Arc. Joan believed in God and worshipped that belief, but her commitment and duty to that belief is what dominated all of her emotions, all of her reason, and all of her intellect; this trust and devotion is the true essence of her character and defines who she was.

Most books written about Joan attempt to apply reason and intellect to her history. Faith and emotion are too abstract, too difficult to measure, too hard to document. On the other hand a more reasoned logic by way of intellect allows for academic assessment, an unemotional recording of the facts. Nonetheless, fact based history is often too cold, too hard, too piercing, too dehumanizing—everything Joan of Arc is not. Thus narrative history was invented, which today is called creative nonfiction, which often romanticizes the facts or embellishes the presumed social activities and human behaviors surrounding particular people or events. While traditional history follows a predictable timeline, creative nonfiction casts off these restrictions to enhance the story wherever deemed necessary to keep the reader interested or commercially entertained. Both techniques have their benefits and faults. For the most part, however, modern professional historians frown on injecting personal beliefs or emotions into their narratives. Michelet has no such constraints. In fact his work attempts to capture an emotional history, the emotions of a faith, the emotions of a belief system, the emotions that brought a devotion to that faith and belief system that is Joan of Arc. By doing so, he has also captured the beliefs of many biographers.

Because of Joan's faith in her faith, she was put on trial by the Catholic Church in the year 1431 for heresy. In her youth she experienced visions and heard voices telling her she had been chosen by God to relieve the English military blockade of the French city of Orleans and then lead the French king, Charles VII, to his official coronation in the city of Reims. Her faith in God created a trust in that message which led to the dutiful accomplishment of her saintly mission, to the great distress of the English who were then controlling large portions of northern and western France at the time—to include the city of Reims. When asked at trial why she believed in her visions and voices, she replied very simply and with a piercing sense of self-perception that it was because she had "the will to believe."[4] She consciously placed a deliberate trust in her faith and then willfully resolved to act on that trust. This deliberate self-control is the mystery we are after and this "will to believe" both by Joan and by those who came to believe in her, will be a consistent thread spinning its elusive web throughout most of our leadership considerations.

Clearly, Joan understood that her heresy trial was a contest between reasoned doctrines and emotional belief, between intellectual Church rules and individual faith, between dictatorial precepts and religious freedom, and she defended her case as much with intellect as with emotion, as much with reason as with faith—a remarkable juxtaposition and combination of paradoxical character traits. So if we ask ourselves what makes Joan of Arc so universal, so seductive, and so relevant to every generation, the answer has to include her faith, emotions, reason, and intellect.

By example, let us jump to the first sentence in Michelet's *Joan of Arc*, which is taken from his book, *History of France*. It reads: "Joan's eminent originality was her common sense." As a measure of Michelet's influence on Joan's recorded history, this idea has already been repeated 164 times

---

[4] Murray, 86.

in just fifty-seven different biographies. But does this sentence really make sense? "Originality" is an odd word to use. To be original or unique is to be "uncommon," while common sense derives from the common customs, conventions, standards, and expectations practiced by a specific culture. But if Joan's common sense was original or unique, then she had to be uncommon, in fact so uncommon as to be the first person or one of the very few people to exhibit common sense. Now plenty of traditional historians would reason against this character assessment *intellectually* by bringing forth any number of prior personalities in history who certainly can be characterized or demonstrated as having common sense. In short, there is really nothing "original" in "common sense," although we may judge it unique.

"Joan's eminent originality was her common sense."

We also know *intellectually* that another obscure definition for "originality" is the "ability to think or act in an independent, individual manner," but when this definition is applied to the "common sense" in Michelet's sentence, we instantly recognize intellectually that all of us possess this same ability to act or think independently or individually. At least we think we do, so there is nothing original or distinguishing there either.

But Michelet is not appealing to our intellect or reason. Instead, he is after our emotions and faith. We understand instinctively what he means, and by briefly suspending our reason and intellect we accept his observation and without objection. Intuitively, we see Joan as "normal" or rational despite her obvious and acknowledged uniqueness. We see her as like us, since most of us see ourselves in this light of common sense—maybe with a few shadows of exception here and there. And because she is like us she is universal, which makes her relevant. We feel this sentence, its emotion, more than we reason it.

By subconsciously normalizing an abnormal phenomenon we become one with the subject. Moreover, we instinctively put our faith in Michelet's judgment, as have so many biographers, knowing full well that *intellectually* common sense is as variable and volatile as any other personality characteristic in the human race. Thus, as recognized by Rudler, Michelet, even with his very first sentence, brilliantly traps us into an emotional and faith based acceptance of history. We instantly drop out of our brains and into our hearts where cold hard facts have little relevance. It is a brilliant opening sentence and we are skewered by Cupid's arrow even without recognizing it.

Having joined Joan to the common sense of the common masses—to us—Michelet's second sentence abruptly separates her from that portion of the common masses who lack *intellect,*[5] who are infected by the "dark contagious force of unreason."[6] Joan, like us, possesses the *intellect* and *reason* that brings "a clear light,"[7] arriving not from ignorance, the lack of intellect and reason, but from the heart—from *emotion* and *faith.* Those of "little faith"[8] could not untie the knot of their times, but with a more reasoned faith and with emotions from the heart Joan could cut through this knot. Here, of course, Michelet is referring to none other than Alexander the Great who, when confronted with the legendary Gordian knot, which if unraveled would reveal the future king of all Asia, pulled out his sword and simply cut the knot in half, solving the problem quite easily. With this analogy Michelet takes Joan out of the ordinary and into the realm of greatness and we feel this greatness through the "clear light" of Alexander's intellect, and through the common sense of the common masses. Common sense because we too believe we would have cut that knot in the same way had we been given the same opportunity. This of course is probably not true, but our egos or emotions tell us otherwise whether we recognize that response in ourselves or not. Yet again Michelet joins us to Joan.

After using Rudler's paragraph to open his introduction to Michelet's work, the book's translator, Albert Guerard, makes the case for the value of *emotion* in historical writing despite occasional factual errors, academic failings, and intellectual vulnerabilities. He argues that to feel an era and to feel a personality is as important as knowing the facts of that era or person and he promises an emotional response when reading Michelet if we but briefly suspend our reason and intellect. "Ignore the factual errors," he tells us, "it's all true anyway. Corrections can always come later." In other words, you can be objectively wrong, yet subjectively right. By fact, despite a treasure trove of records and documents about Joan of Arc's life, most of her history is still in debate 600 years later. Objective facts may not be wrong, but the many intellectual interpretations subjectively derived

[5] "This sets her apart from the multitude of enthusiasts who, in ages of ignorance, have swayed the masses." Michelet, 3.

[6] "In most cases, they derived their power from some dark contagious force of unreason." Ibid.

[7] "Her influence, on the contrary, was due to the clear light she was able to throw upon an obscure situation. . . ." Ibid.

[8] "The shrewd and the cautious, the men of little faith, could not unravel the knot: she cut it." Ibid.

from these facts still suffer from logic failures, especially where it concerns Joan's military operations with medieval armies, as will be addressed later.

Of course relying on emotions can lead us astray, especially when applied to judgments and decision-making. Style makes us feel good, but can be deceptive. Substance on the other hand can feel bad, disappointing, even make us delusional or provoke self-denial. Facts disrupt the synapses of our brain while emotions glide smoothly along pre-channeled, well-worn paths. It is easier to be politically correct than to gather up arms against the prevailing attitudes of the times. In other words, it's easier to feel than to think, to go along with things than to fight against them, to follow the lemmings over a cliff rather than grab the edge and yelp for help. When the philosopher Schopenhauer asked,"What is it that forces the man...to think and act with his herd, and not seek his own joy?" his answer was because of "idleness, taking things easily."[9] Nevertheless, to be both emotional and substantive will be the intent of our study which is another reason for choosing Michelet's writing for comparison, as opposed to the hundreds of other Joan of Arc biographies.

In another example of Michelet's emotional appeal, the first sentence of his second paragraph reads "It was not rare for women to take up arms."[10] Now think about that statement for a moment. Were women really everywhere on the battlefields of medieval history? Certainly there were women everywhere within medieval armies but almost exclusively as camp followers—wives, servants, prostitutes—not with swords, knives, mace, and lance. Do we really see women as armored knights on horseback? Do we really see them amassed and aligned on the battlefield ready to meet an advancing enemy mixed with men and more women? Do we really imagine them in hand-to-hand combat? Michelet cites a few examples of such cases as if they were common to history, but factually we know they were not common—in fact they were rare, extremely rare. By analogy, we know that some tigers in the jungle are white, but that does not make all or even many tigers in the jungle white. Similarly, just because some very small, minute number of women participated in medieval combat, not all or even many women participated in combat. Instinctively we know his syllogism to be false.

"It was not rare for women to take up arms."

[9] Nietzsche, Friedrich, *Thoughts Out of Season*, "Schopenhauer as Educator." Adrian Collins, trans., 104.

[10] Michelet, 3.

However, when we *feel* Michelet's sentence we get a different impression, an emotional translation, the sensation that—well—women can and have fought on the battlefields of history and therefore we can extend that fact to embrace larger numbers. To believe this, however, we must again suspend our *intellect* and place our *faith* in Michelet's judgment. In this regard, we are left with the impression again that Joan was common, average, middle-of-the-road, a normal occurrence, and...like us. Like us, because we want to believe—we *will* to believe—that women can fight on the battlefield, and because we will our mind to accept that belief, especially in these modern times, we place our *faith* in that belief, even though we know intellectually that belief is false concerning medieval women. Even before Christ, Caesar recognized in his *War Commentaries*, "Certainly we are prone to believing what we want to, and rather hope that others *feel* what we feel ourselves."[11] Michelet is trusting in this ancient human frailty when he exaggerates his case to achieve an emotional compatibility between his subject and readers. These examples also explain the *reasoning* behind how Joan of Arc functioned personally within the world around her, hoping others would feel what she felt in her devotion to her faith.

Joan of Arc, however, was not only rare, she was so rare as to be one-of-a-kind, a unique gem mined only once from all of the quarries in human history so far. Mark Twain described her as "easily and by far the most extraordinary person the human race has ever produced." Now that is some exalted praise and certainly takes her out of the common. But does it eliminate Michelet's emotional appeal, the idea that all of us fit into her categorization?

As we know, despite our commonality as a species, every fingerprint is unique and that makes all of us extraordinary, a marvelous creation never to be repeated within the human race. And cannot women fight our battles if called upon to do so? To understand Joan of Arc—her personality—and to believe in her, we somehow have to take her out of that surreal world of exceptions and place her into the real world of our acceptance. Michelet creates that descent by using emotional history, otherwise our intellect would reject it.

Continuing in this vein, Michelet begins his third paragraph with the sentence, "Neither is the uniqueness of the Maid to be sought in her visions." Really? Again he cites a few examples of supposedly ubiquitous visionaries

[11] Warner, trans., 268.

hovering everywhere about the royal courts of medieval Europe, which idea is copied endlessly in other Joan biographies.[12] But listing a short series of female mystics, often decades apart and stretching through centuries, claiming they were common to the times and thereby of influence concerning either Joan of Arc or Charles VII, does not make sense, even for the medieval mind. Often Saint Bridget of Sweden who died in 1373 or Marie de Avignon who died in 1399 are mentioned, both of whom lived decades before Joan emerged in 1429. But these women, as with a few others, were all separate of each other, not influencing each other. Nor were the social and cultural conditions so uniform within Europe or even within France as to breed visionaries equally alike from both urban and rural communities across a wide spectrum of diverse geographical regions. Joan did not see visions because others saw visions, nor at the age of twelve or thirteen was she likely aware of any contemporary visionaries when her own visions first appeared. Besides, despite her best efforts and rarely mentioned in Joan biographies, Marie de Avignon was unsuccessful in her attempts to meet with the French king and had to "settle for an interview"[13] with the king's wife. Royal access, as eventually achieved by Joan of Arc, was not as easy as some biographers like to claim.

"Neither is the uniqueness of the Maid to be sought in her visions."

Some argue that an all-male excessively religious medieval culture forced the emergence of various aberrant female behaviors, one of which was mysticism, in an effort by women to express equally the magic produced by male priests at the altar; that is, transubstantiation, the changing of bread and wine into the body and blood of Christ.[14] This concept also leads to the idea that in their efforts to produce a comparable magic women came to rely on folklore, which reliance led to the magic of visionaries as well as to witchcraft, or at least to perceptions of witchcraft. Here

[12] "when the air is infected [with visionaries], everybody breathes the contagion." Lamartine, 128. "Clairvoyants and wizards [were as] common as beggars and lice." Waldman, 23. Goldstone, 93, has them in every "hamlet, town or village in fifteenth century France." See also Michelet, *Joan of Arc*, 4, and Parr, I, 6. Thus, since Joan never revealed her visions while living in her village, there must have been more than one visionary in Domremy. It also seems that others in Vaucouleurs, Nancy, Neufchateau, etc, must have also failed to reveal their visions since they supposedly existed everywhere. Or are we creating our own myths?

[13] Hobbins, trans., 30.

[14] For example, see Barstow, 42.

then we have folklore becoming folk magic mutating into a folk religion, which allegedly became Joan's unorthodoxy—feeding "her belief in her own powers."[15] Michelet fuels this idea when he says, "Her religion came to her, not as a lesson or ceremony, but in the folk like and naive form of a lovely tale told of an evening, like the simple faith of a mother"; and then later, "Unawares, the young girl created, so to speak, her own ideas, turned them into realities, made them entities, powers, imparted to them, from the treasure of her virginal life, an existence so splendid, so compelling that the paltry realities of this world grew faint by comparison"[16]—all great style, but with virtually no substance. Her mother taught Joan her basic prayers, but to what extent Joan knew Bible stories or French legends and where this knowledge came from—on "an evening"—is not known. Here we have creative nonfiction. That Joan was delusional is pure conjecture. To date, no one has been able to explain convincingly why Joan of Arc saw visions, nor are we likely to ever know why.

The logic problem in all of this is that practicing the rites of a regional folklore does not incur a cultural belief in that folklore, just as the display of a Christmas tree does not give evidence of a literal belief in Santa Claus. Knowing folklore and practicing the rituals of folklore does not imply a belief in the characters or powers within that folklore, else we would have every kid in America putting on capes and jumping off rooftops. Missed by Michelet is the fact that for folklore to become a religion, religious belief implies worship more than ritual, outcomes more than custom, and the promise of salvation more than protection or healing, which neither magic nor folklore ultimately provides. And where you have worship, outcomes, and especially salvation, you have more than delusion—unless of course one abides by the cynical reasoning that all religion is artificial and nothing more than a metamorphosis of folklore.

Dancing around a tree during holiday celebrations in Domremy does not imply worship; knowledge of mandrakes fomenting riches does not imply a commensurate outcome; and the symbolic practice of using amulets, which goes back as far as Sumeria 5,500 years ago, does not imply salvation. Most likely more visionaries were imprisoned or burned at the stake than lived to reach the pages of history. Joan was not accepted into the royal court of France because visionaries were common nor because royal courts routinely embraced them.

---

[15] Barstow, 53.
[16] Michelet, 8,9.

Moreover, the idea that the mere existence of multiple visionaries creates a consistent belief in their credibility and a general acceptance and reliability in their prognostications is also an irrational logic, even for the medieval mind. By example, just because people in our own era see UFOs, and just because half the people believe in their existence, does not mean that a person who sees a UFO will be admitted to the White House and put in charge of the nation's air defenses, whereas Joan of Arc was given royal access and put in charge of her nation's armed forces. Further, a government may accept reports of these sightings, but this does not imply an official belief in or proof of extraterrestrial existence, while the majority of the population may confess and accept that belief. Further, a president may be forced to investigate this popular UFO phenomenon, just as a king may be forced to give audience to an alleged visionary, but these acts are not a validation or proof of beliefs. Consulting, investigating, and exploring are different from accepting, implementing, and trusting.

But Michelet would have us believe this was the case, making Joan once again "common," just one of many visionaries, while still injecting greatness into her character. The mystic personalities named by Michelet—Friar Richard, Marie de Avignon, Catherine of Rochelle, and the shepherd boy—four people in a population of twelve million, is hardly ubiquitous. Would anyone have even known of these charlatans had Joan not existed so they could be included in her history? Even if we accept Michelet's logic, one Joan biographer made the enlightened statement, "It does not matter to say that Shakespeare was one of a group, that he was an Elizabethan dramatist. The real point is that he was very different from all other Elizabethan dramatists."[17] It may be that Joan of Arc was the spiritual Shakespeare of our times. Meanwhile Michelet tugs us toward both ends of the historical spectrum; Joan is common because she is uncommon.

Michelet goes on to discuss the geographical character traits of the rural culture surrounding and forming Joan's belief systems. Once again, he begins at the extreme edges and brings us back to the core. The adversarial province of Lorraine, just across the Meuse River from Domremy, did not produce Joan of Arc, as Michelet eventually goes on to explain, but

[17] James, 94. Cristiani, 28,32, also points out that Joan was different from other visionaries because (1) she had so many visions—he estimates the number somewhere between five hundred and seven hundred and fifty—and because (2) she then kept them secret for several years unlike most visionaries. This information, however, could not have been known by Charles VII or his advisors. Gies, *Joan of Arc*, 25, points out by historic example that Joan was different because most mystics were sickly whereas Joan was the picture of good health.

The Meuse River border between Champagne and Lorraine.
(Image@2016 Google Earth)

he insinuates its influence and draws a sharp if somewhat fanciful contrast between the Lorraine character on one side of the border and that of Joan's province of Champagne on the other side, reminding us there are always two sides to every border. By fact, after poisoning our minds with the Lorraine character, we find she is really from the Duchy of Bar, more appropriately a part of the province and character of Champagne. Thus she becomes both "wily and crafty,"[18] genuine and modest, a blend of two regional characteristics, a mixture of emotions and reason, intellect and faith—like us.

Michelet tells of the border wars along the Meuse River that came before and after Joan, as if past history and future history permeated every peasant farmer in medieval France. Again, intellectually, we know this cannot be true. He suggests that military raids on or near Domremy were so common as to keep the peasant farmers constantly on alert and by this means Joan came to know war. This of course is emotionally effective, but patently untrue again. By analogy, we might say that just because America practiced civil defense measures during the Cold War, such as designating protective shelters, hiding indoors when radiation clouds passed overhead, testing sirens, and learning to duck and cover, all American children knew nuclear war. We know this is untrue, but emotionally we accept Michelet's

[18] Michelet, 4.

premise anyway. There was a war on that surrounded Joan in her childhood and it would have had some psychological effect, but to what extent is only conjecture. No one from Domremy, except Joan of Arc, ever died in what is known to history as the Hundred Years War between France and England, a war that allegedly began in 1337 and ended 116 years later in 1453, twenty-two years after Joan's death. In fact, Joan would not know war until she reached the battlefields of France and then she would know it only too well.

Thus a brief interlude is warranted to survey the actual conditions existing inside France during Joan's era to establish more appropriately our feel for the real environment in which she grew up and operated, and which influenced the development of her leadership talents. We may be surprised at some of the similarities to our own times.

Many modern historians and not some few biographers seem virtually possessed with an incurable mania for depicting the most outlandish martial and social circumstances inside France at this time. Much of their hysteria comes from such sources as the Bishop of Mans, Martin Berruyer, who in 1456 described early fifteenth-century northern France as if from Dante's inferno. Virtually no crime or condition escaped his accusations.[19]

Certainly medieval France was no bed of roses, but it remains critically important to separate normal community hardships from those unique to the times, especially those unique to existing political and military circumstances, allegedly generating the need for Joan's intervention in French history. For example, destructive insects and damaging weather occasionally ruined harvests, and deadly epidemics periodically swept through the population, but these disasters cannot be attributed to the Hundred Years War or to its political intrigues as so often alleged. Normal medieval life was fraught with peril, but then so too is normal modern life if we but select our data with the same bias.

"It remains critically important to separate normal community hardships from those unique to the times."

The *Journal of Paris*, written anonymously during Joan's era, offers considerable insights in this regard. By medieval standards Paris was a large city with a population of 100,000, yet in 1418 when Joan was just six

[19] Lynch, 55; Wyndham, 52-3.

years old, it suffered 80,000 deaths from the plague,[20] a ghastly experience. Bodies were stacked in piles thirty to forty deep inside expansive burial pits.[21] Again, twenty years later in 1438, after rural migration repopulated the city, Paris suffered another 45,000 deaths in a single epidemic.[22] In 1426 and again in 1443 whooping cough arrived, and in 1432 smallpox and plague attacked the city, which suggests epidemic cycles every six to eight years.[23] In fact by 1450 the nearby "population of Normandy was reduced to less than a third its size before the Black Death...Many villages were deserted."[24] These were not unlike the epidemic conditions found among the native American Indian tribes of the Pacific northwest during the Lewis and Clark expedition 400 years later.

It should come as no surprise then to find the anonymous bourgeois author writing in his *Journal of Paris* at the time, "How very many places are now uninhabited, towns, castles, churches, abbeys, and other places beside." But this author, who appears to have traveled no more than a few miles from Paris in his entire lifetime, mistakenly follows with "Alas, Alas, how many orphans there are now in Christian lands, how many women left widowed and helpless by all this killing," as if wars or crimes were causing his observations instead of urban epidemics and agricultural famines. He then goes on to decry the high rents demanded in 1424 from the landlords of the city and declares, "you would have found more than twenty-four thousand houses in Paris in good repair, perfectly sound and solid, standing without a soul living in them." Greedy landlords are common to every age and those in Paris were refusing to lower their rents, creating a homeless population, also a common modern problem.[25] It never seems to cross his mind that a single event, war, was not the exclusive reason for his limited observations.

Moreover, biographers, so quick to recite these lamentations, rarely compare these conditions with those existing outside France at the time.

[20] Haggard, 198; Richards, 47; Paine, I, *Joan of Arc*, 74. Neillands, 112, claims the population was 150,000 and the deaths 50,000. Leary, 12,67, has the population at 200,000.

[21] Richards, ibid.

[22] Pernoud, *Joan of Arc: By Herself and Her Witnesses*, 257, says 50,000 deaths.

[23] See also Gottfried, 133. By historical example, "Major disasters like fire, earthquake, famine or epidemics ravaged the major city of Antioch once every 15 years during the 600 years of Roman rule." Wade, 161.

[24] Pollard, 30-1.

[25] As an example of the universality of this issue, during Mao Tse-tung's rise to power in China he organized rent strikes and anti-landlord demonstrations among 2,000,000 peasants of Hunan. Tuchman, *Stilwell and the American Experience in China 1911-45*, 132.

For example, Barcelona, Spain, suffered epidemics in 1396, 1410, 1429, 1439, and 1448, while also suffering from frequent crop failures and military conflicts.[26] Its population decreased from 50,000 in 1340 to 20,000 in 1477,[27] appearing to suffer in the same manner as Paris. When Pope Martin V arrived in Rome in 1420 "he found the city...so ruinated that it looked not like a city. You might have seen the houses ready to totter, and churches fallen down, the streets empty, the city full of dirt and mire, and in extreme want of all sorts of provisions."[28] Clearly, fifteenth century Europe suffered from urban decay and decreasing populations, and war was not the exclusive reason for these problems.

But this general lamenting, which historians accept so literally, seems to be a constant ingredient in the cultural soup for medieval France. Compare the *Journal of Paris* with these comments from Petrarch in 1360, seventy years before Joan of Arc:

> Everywhere is solitude, ruins, misery; the fields are deserted, the houses dilapidated and empty, except in walled towns; everywhere you see...scars still bleeding from English swords [maybe from the devastating Battle of Crecy that had taken place four years earlier in 1356]...Where are the crowds of students, the colleges overflowing, the great riches?....In place of philosophy, of literary discourse, of peaceful conversation, you hear the cries of arms, the bellowing of battering rams beating the walls, and the alarm of soldiers fills the streets [of Paris] with tumult and fear![29]

Again, soldiers are blamed for conditions clearly derived from Bubonic plague, the Black Death, which killed from one-third to one-half the European population between 1347 and 1349, just a decade before Petrarch's quoted observations. If anything was truly unique about medieval life, it was the constant dangers posed by epidemics, not by war. And these standard lamentations continued to salt French tears even up to Joan's time.

An awareness of these plagues and their devastation is necessary to one purpose: to acknowledge the biological chaos immediately preceding Joan of Arc—atrocities greatly exceeding even the percentages of world losses in our own modern wars, to include two atomic bombs. These epidemics left

[26] See also Gottfried, 133, who cites eleven epidemics in Barcelona between 1396 and 1437.
[27] Bisson, 165.
[28] See Barrett, 492, and Brown, 477, who quotes from Platina, B. *The Lives of the Popes*. London: 1685.
[29] See also Tuchman, Barbara W. *A Distant Mirror: The Calamitous 14th Century*, 198.

behind a community of survivors, a hardy stock of resilient people, possessed of a healthy respect for the unpredictable. Here was a population that lived but knew not why, except by the grace of God. Joan and her audience were of this legacy. If God could take, then God could deliver, and the mysteries of death could be equally applied to the mysteries of life. In this way the medieval heart knew more of the spirit[30] than do our own minds of prosperity whose science is yet to catch up, which goes a long way toward explaining Joan of Arc.

Crop failures were another normal hazard. Freezing weather, floods, or severe hailstorms, as happened in 1423 and 1427, could wipe out budding fruit trees, destroy vegetable crops, and disrupt the flow of goods and produce, which lacked the benefits of refrigeration. Food arrived fresh from the fields each day or not at all.[31] Pesticides were unknown and caterpillars frequently destroyed fruit crops while moles periodically ingested ground harvests, as happened in 1425, 1428, and 1440. Certainly these events were disasters, but they were not extraordinary or unique to France at the time, especially as concerned the need for Joan of Arc to save a "suffering" country.

We should not be surprised then to learn that Joan most often consumed only one meal a day, never more than two, and her standard fare was bread dipped in well water purified with wine,[32] a practice going back a thousand years to ancient Greece where they first began mixing three parts of water with two parts of wine.[33] Joan may have acquired this liquid habit from eating hard bread, suggesting the common use of some form of hardtack or crackers, long a soldier's staple. The early crusaders invented and carried crackers, a twice-baked bread, because they survived storage by resisting mold. Joan's eating habits may also suggest she had dental problems, since her diet seemed short on fruits and vegetables, resulting in a possible mild form of scurvy, a condition suffered by both Philip Augustus and Richard the Lionheart during the Third Crusade.[34]

---

[30] See also Taylor, *Joan of Arc: La Pucelle*, 31.

[31] Stores of food included pickled fish, salted or smoked meat, and sun dried fruits and vegetables. Gies, *Life in a Medieval City*, 50.

[32] Murray, 262. Testimony of de Contes. Also in Parr, I, 136; Paine, I, *Joan of Arc*, 157,175 (*l'eau rougie* is water reddened with wine); Reid, 18,48.

[33] Durant, Will. *The Life of Greece: Being a history of Greek civilization from the beginnings, and of civilization in the Near East from the death of Alexander, to the Roman conquest; with an introduction on the prehistoric culture of Crete*, 296.

[34] Reston, 193.

But there were good years also when cabbage, peas, beans, and fruit were plentiful as in 1423; wine was excellent and cheap in 1424; herring, cherries, plums, and almonds were profligate in 1426; beer and eggs abounded in 1428, and so on. It was not an ideal life, it had its ups and downs, but the problems of France were clearly not caused exclusively by the various warring factions. Nonetheless war did add to these frustrations and contribute to this mixture of regional problems.

To help finance the Hundred Years War for both England and France, monetary currency was occasionally devalued, as in France in 1423 and 1424, and heavy taxes were sometimes levied, as in France in 1423 and 1427, which certainly compounded the hardships, especially for the poor. An Englishman complained, "Alas! When a poor man shall have paid his impost, his villein tax, his salt tax, his hearth money, his fourth, the King's spurs, the Queen's belt, the customs, the road tax, the tolls—not much remains to him. Then will come another levy newly created."[35] Have you looked at your modern telephone bill lately? Of course, as with Saladin's Tithe,[36] a ten percent tax to support the crusades, knight and priestly accoutrements were often exempted from these tax levies adding to the fiscal pressures on the general populace. Similar tax exemptions of course exist in modern times. These medieval taxes were commonly paid with farm products, bushels of wheat, geese or pigs, and sometimes with labor rather than with coinage. But these domestic tariffs are natural expectations for most political emergencies in early history, if not in all times where periodic excessive taxation and increased inflation devalue the people's currency.

The city of Paris also suffered from crime, which seems normal to the human race. However, the only abnormalities about these crimes were their occasional consolidation, as with our modern mafias and cartels, as happened outside Paris in 1423 and 1426. But in those years the Provost rounded up and jailed fifty and two hundred criminals, respectively. It's not as if civil authorities were not policing the problem. In fact, it is alleged that the duke of Bedford hung 10,000 brigands in 1428 alone.[37]

Further, packs of wolves periodically entered the city for brief trials, usually digging up shallow graves, killing off domestic pets, or chasing farm animals, but also occasionally attacking humans, which of course

---

[35] Wolf, 233n, quoting from Gerson, in Fr. Funck-Brentano, *The Middle Ages*, 498.

[36] Billings, 106.

[37] Durant, *The Reformation*, 70; Smith, 23.

historians make much of in their hysteria.[38] But these occurrences were no more frequent than shark bites off the coast of Florida today. In fact, throughout her childhood, Joan frequently traveled alone through deep woods just to visit the religious shrine of Notre Dame de Bermont, apparently without fear of wolves. There is a big difference between the hysteria of urban unfamiliarity with the wild and practical peasant rural reactions to these same dangers. On the other hand civilian homicides, rapes, and mayhem are rarely mentioned by this Parisian chronicler.

But the real problem with describing the various regions destined to become greater France is that the Paris region was not the whole of France. The provinces of Burgundy, Brittany, Normandy, Maine, Provence, Lorraine, Champagne, et al., experienced a variety of economic and political conditions and not all were in dire straits: some dukes were generous and caring, others brutal and exploitive; some churches were charitable, others parasitic; some markets were constantly filled, others sold rotten remnants; some beggars starved, others were cared for by the populace at large—conditions not unlike the differences that sometimes exist between our various American states. Highway robbery was a threat, but then so too were pickpockets in Times Square.

There were of course feuds between the various nobles, mostly over property rights as a result of England's attempts to enforce the Treaty of Troyes which gave control over all of France to England in 1420. Occasionally gangs of hired thugs would venture forth to enforce property claims based on this treaty.[39] This, however, resulted mostly in the theft of cattle, some minor arson, and a lot of kidnapping and ransoms, but not in a lot of killing.[40] Moreover, these forays were random and rarely sustained for any length of time. When grouped together over several years, however, they can be made to appear more prevalent than they actually were. As Louis Bertrand observed, "France in those days was not then so utterly disordered, so completely at the mercy of the soldiery, as one [Martin Berruyer?] would have us believe."[41]

---

[38] See Eliot, 175-76, Haggard, 199, Winwar, 16, Durant, *The Renaissance*, 30, and Maxwell, 128. The *Journal of Paris* reports that during a one week period fourteen people were killed by wolves in and around Paris, but once the pack leader was killed these incidents ceased. James, 189, and Ross, James Bruce and Mary Martin McLaughlin, eds. *The Portable Medieval Reader*, 228.

[39] See Marot, 18-23.

[40] Marot, 20-2.

[41] Louis Bertrand, "Joan in Lorraine" in *For Joan of Arc*, 20.

Thus, another problem derived from this period in French history might be called provincial transference; that is, conditions that dominated in one French province were often transferred by medieval historians onto other provinces,[42] and then made to include the whole of France by modern historians, not unlike claiming that all of America is like California or New York. This seems particularly true concerning the problem of brigands. English-occupied Normandy in *western* France had a pervasive history of outlaw bands and this localized infestation is often transferred inappropriately to other French provinces, to include Joan's province of Champagne in *eastern* France.[43]

"Another problem derived from this period in French history might be called provincial transference."

Certainly life, especially the good life, was not a sure thing, but then it never is. The idea that social and economic France was worse off in 1429, when Joan emerged, then say in 1360 or 1423, is a contrivance of academia and not of much value in studying Joan or her times. "Thus," as Sackville-West said in her popular book about Joan of Arc, "it is probably as well not to exaggerate, as some historians would seem to have exaggerated, the troubles suffered by the peasants of a little village like Domremy."[44]

Yet historians like Michelet do exaggerate these conditions, especially as they relate to war. Maurice Keen, an award winning British historian specializing in the Middle Ages, even goes so far as to state, "[the passage of armies] made the ravages of war into a factor of social and economic importance at least comparable with the effect of the plague."[45] Really! Warring soldiers were as prevalent and virulent as the multitude of fleas and rats bringing the epidemics of Bubonic plague?[46] The minor destruction and death tolls in battle, especially compared to modern times,[47] equaled

[42] "the medieval thinkers and artists were 'great provincials', unable to outgrow their provincial standards and see further than their church bells sounded." Gurevich, 68.

[43] Endore, 415; Pollard, 31; Adams, ix.

[44] Sackville-West, 40.

[45] Keen, 228.

[46] Even during the disastrous and bloody Third Crusade, disease and famine took more lives than did combat. Reston, 336.

[47] During World War II, there was an average of 27,600 deaths every day, "or 1,150 an hour, or 19 a minute or one death every three seconds." Atkinson, Rick. *An Army at Dawn: The War in North Africa, 1942-1943*, 5. On the other hand, the Black Death, Bubonic plague, killed an average of 22,000 a day.

the loss of one-third to one-half of Europe's population? The sometimes criminal activities of looting, kidnapping, and spoiling were as economically and socially devastating as those resulting from the traumas of heavily depleted populations and the multitudes of unattended fields which caused the scourges of agricultural famine? If so, between armies, plagues, and starvation there would have been nobody left. Again, it is "hysterical" history.

There are some few who agree. Guy Endore, in his book about Joan, is particularly appealing in his observations. He writes:

> Personally I am inclined to think that our opinion of that period (Gothic) is like the New Yorker's opinion of Chicago,[48] or the European's opinion of the United States.
>
> We must ask ourselves how many of the epithets we so glibly apply to the Middle Ages can be applied equally well to our own age. The cruelty? Have we ceased to be cruel? Our wars, are they less bloody?... Was there no charity then, no kindness? Or do we think these modern inventions?
>
> Have we not our grafters...Our jealous generals and our traitors? If we don't have [Joan of Arcs] it is perhaps because instead of sending them to commissions of religious men to be examined, we bundle them off to insane asylums or take them away to hospitals to operate on them.
>
> It is true that they imprisoned men for opposing the official religion, while we only imprison those who oppose the official financial and social order. They had to use faggots on [Joan]. We know how to burn with electricity. The advance is obvious.
>
> People no longer starve in the United States. They are only undernourished. By the millions are the half-starved counted. Yes, famines are a thing of the past. Unemployment is its new name.
>
> No longer do we need castles on every hill. We need only battleships, burglar alarms, fortresses, policemen, subterranean bank vaults, etc., etc.
>
> No longer do robber barons extract tolls from every peaceful merchant. That is now the business of the racketeer.[49]

[48] Chicago alone experienced more than 760 homicides in 2016.
[49] Extracts from Endore, 325-27.

Domremy France (Image@2016 Google Earth)

Of course Endore was writing in 1931, in the face of the Great Depression, but his complaints still ring true. The hundreds of millions killed in wars, crimes, famines, and epidemics even since his time would certainly leave his pen dry today, and the jealousies and graft continue. We can only imagine what historians might write about us six hundred years from now.

Even more bothersome is the often persistent translation of this hysterical history directly onto Joan's village, which seems to have been rather isolated from it all. The idea from Michelet that she "realized the full meaning of war" or that "she knew first-hand the horrors of war"[50] while growing up, as echoed in another recent book about Joan, is hardly credible. Simeon Luce, whose book on Joan perpetuates many of her current myths, was even worse in his hysteria, describing a constant string of devastating raids on Domremy.[51] Yet by all accounts, no Domremy villager was ever killed or captured, and the village livestock survived the only raid that ever reached the town. Otherwise the recorded losses were some burnt homes and a destroyed church during which the village population of some 100 people, according to a 1398 census,[52] nestled itself safely away inside the nearby town of Neufchateau. Joan was hardly preconditioned to the human slaughter, death in the thousands, eventually experienced on the battlefields of France, the true horrors of war.

---

[50] Taylor, *The Virgin Warrior*, xxi.

[51] Endore, 415.

[52] Marot, 15.

When Joan said to the priest Jean Lombard, during her interrogation at Poitiers, "God has taken pity on the people of France," an often quoted phrase, she did not mean so much the actual people as the monarchy itself. In fact if we take into account the frailties of translators, Albert Guerard, the translator of Michelet into English, renders this same passage as "the great *pity* that was *in* the kingdom of France."[53] Now pity *on* the people and pity *in* the kingdom are very different perspectives. To Joan, as to many others of that time, the great pity in the kingdom was Charles VII. His suffering was the people's suffering.[54] At trial, Joan was asked, "Had you in your youth any intentions of fighting the Burgundians?"—French rebels from the province of Burgundy. Joan's answer was very specific and should leave absolutely no doubt as to her intentions, "I had a great *will* and desire that my King should have his own Kingdom."[55] Her motives cannot be stated more clearly, yet historians and biographers persist in the idea of placing Joan on a religious crusade to rescue France from all its perceived economic and social sufferings,[56] which simply was not the case. Others claim that her reference to pity meant the suffering of the people undergoing the blockade of Orleans, and that Joan's only mission was to relieve that city; crowning the king was an afterthought,[57] which idea again conflicts with Joan's own testimony, and which idea again demonstrates the subjective nature of intellectual reasoning when derived from objective facts. How some authors come to their conclusions is often as mysterious as Joan's own visions.

"I had a great will and desire that my King should have his own Kingdom."

The English were occupying the French king's rightful inheritance and Burgundian rebels were aiding in this larceny. Others were extorting or bribing to take their share of his royal domain,[58] seeming to rob Charles VII and thereby the French people of their wealth—paid in taxes. English

[53] Michelet, xi. See also Parr, I, 25, II, 130, and Lynch, 23,81.

[54] Royden, xiv, seems to agree: "Charles was not Charles to her, but France…France was embodied in her King." Vale, 194, amplified: "The king was not one, but two persons: he was the God-made man, destined to perish; and he was the man-made God, personifying the 'mystical body' of the kingdom, which was immortal."

[55] Murray, 19.

[56] For example, see Smith, 46, Endore, 14, etc.

[57] For example, see Spoto, 56.

[58] Lowell 9; Endore, 28.

rumors had made Charles VII a laughing-stock and his royal authority appeared to be in shambles. It was this terrible state of royal affairs that maligned the realm, prodding Joan to her mission, rather than the common plight of the people who were confused, disheartened, and occasionally desperate, but otherwise the same. Joan of Arc, by crowning Charles as the legitimate king of France, hoped to weaken this English claim to the French throne and thereby dilute their motives for occupying France, alleviating the king's royal suffering and ensuring his personal safety.[59] Her primary mission therefore was more political than humanitarian, and her military mission to relieve the blockade of Orleans was necessary only to achieve that political goal.

Although medieval romantics, and some many modern historians, bemoan this collective human condition in France at the time and invariably include it as an all-encompassing natural addendum to their perceptions of French history, no one, including Joan, could have reasonably believed that Charles and his coronation would improve this day-to-day life for the French people.[60] More likely it might simply accomplish nothing more than provide for a closer and more sanctioned channel of civil complaint in the person of Charles VII who could be more readily accessed and influenced than could an alien English monarch isolated on an island. Nonetheless, it might also lead to winning the war, lowering taxes, and unifying a divided people, which political vision seems to have been exclusive to Joan—or to her God.

Here then we have more realistically set the environmental conditions and the social and economic stresses that contributed to Joan's character and leadership. Thus, with Guerard's introduction to Michelet's *Joan of Arc*, and with Michelet's first few pages, we already begin to feel a blend of contradictions: a world different from ours, yet the same; inhabited by a personality greater than the times, yet of this world; with a soul and spirit unique in all of human existence, yet like us.

After Joan emerged to stun the world with her military exploits, it should come as no surprise that a contemporary of Joan's, the French poet, Alain

[59] "But the Maid was always of the opinion…if once the King were consecrated and crowned, the power of his adversaries would decline, and that in the end they would be past the power of doing any injury, *either to him* or to his kingdom" (italics added). Murray, 239. Testimony of Dunois.

[60] "throughout the Middle Ages there was not one single attempt to get to grips with the root cause of poverty; the poor continued to be fed, the donors gained virtue, and the situation remained as it was." Gurevich, 243.

Chartier, would write of her, "Here is she who seems not to issue from any place on earth" or that Alexander Dumas would call Joan "The Christ of France." After all, there is little difference between a manger in Bethlehem and a thatched hut in Domremy; a poor carpenter and a poor shepherdess; a boy debating Jewish scholars in a temple and a girl debating the Catholic hierarchy in Poitiers or at Rouen; a gathering of disciples of whom not all are loyal; a donkey or a horse for transportation; praying in the garden of Gethsemane or in a vineyard at Orleans; an appearance of invincibility in the face of Roman or English rule; a stake or a cross for crucifixion; resurrection from the dead, perceived or real—and unbelievable events, divine or miraculous. God sent Christ into the world and Joan of Arc declared His same authority, and if God has a Son, why not a Daughter? The plague of Black Death foretold of such comings, the Second Coming,[61] and could not Joan in some fashion represent the deliverance of that promise? Even the more modern Sir Arthur Conan Doyle, famous for his Sherlock Holmes, commented: "my personal conviction [is] that, next to Christ, the highest spiritual being of whom we have any exact record upon this earth is the girl Joan. One would kneel rather than stand in her presence."[62] For the moment we will not pursue this reasoning, but it demonstrates just how powerful this Joan of Arc story can start to become.

So was it really just a mixture of a war ravaged countryside and distinctive regional personality traits, as claimed by Michelet, that created Joan of Arc, or were there other more powerful forces at work, forces that created a profound leader who could lead men in combat?

We have now set the stage for our study by recognizing faith, reason, emotion, and intellect as two sets of competing yet complementary guideposts. We have discussed Michelet's approach to Joan as creative nonfiction and acknowledged the appropriateness of sometimes suspending our reason and intellect when reading the history of Joan of Arc to make room for emotion and faith. We have debunked the myth of ubiquitous mysticism and witchcraft among medieval women and the idea that women routinely served in medieval armies. And, finally, we have challenged the usual medieval stereotypes by recognizing a more normalized, humanized

---

[61] Revelations: 15:1, "And I saw…the seven angels having the seven last plagues: for in them is filled up the wrath of God."

[62] Denis (in the translator's unnumbered preface); See also Nadia Margolis, "The 'Joan Phenomenon' and the French Right," in *Fresh Verdicts on Joan of Arc*, Wheeler and Wood, eds., 269.

version of French society and culture during Joan's era while acknowledging its more unique aspects of frequent epidemics and famine, creating a world of hardy and resilient survivors. Within this framework we can more accurately explore Joan as a leader of men.

Having set the stage for Joan, we can now lift the curtain on the more formative foundations of her leadership as derived from her early environment to examine where these lessons came from and, to an extent, how she used them, and whether we too can acquire these same leadership skills. The answers may be surprising.

Jeanne d'Arc écoutant ses voix (Joan of Arc listening to her voices)

François Rude (French, 1784–1855)

Louvre

Chapter Two

THE FOUNDATIONS OF HER LEADERSHIP

Having set the stage for our study by using Michelet's *Joan of Arc* as a guide and recognizing faith, reason, emotion, and intellect as two sets of competing yet complementary guideposts; acknowledging Michelet's approach to Joan as creative nonfiction, and the necessity to suspend reason and intellect occasionally when reading history to make room for emotion and faith;[1] debunking the myth of ubiquitous mysticism and witchcraft as a religious movement among medieval women; and challenging the usual medieval stereotypes by constructing a more normalized and humanized version of French society and culture, with the exception of epidemics and famines during Joan's era; we can now begin to explore the foundations of her leadership.

Remarkably, Joan as a soldier and military commander is almost universally ignored in most biographies. Only two books attempt to define her role as a military leader and neither explains where her leadership talents came from. In 1999, Kelly De Vries, in his book *Joan of Arc: A Military Leader*, concludes simply that "Somehow, somewhere, sometime, perhaps over a long time, in many places...Joan became a leader."[2] This study defines the somehow, somewhere, and sometime so elusive to De Vries.

In his 2003 book *Joan of Arc: The Warrior Saint*, Stephen Richey briefly credits Joan's parents as a likely influence cultivating her leadership talents and accepts certain family responsibilities as conducive to

[1] "Faith expands the heart and mind, creating that inward spaciousness in which alone reason can work," Anselm in Heer, 105.

[2] De Vries, 56.

that leadership development,[3] but he then drifts off to compare the broader principles of war with Joan's demonstrated use of them, which may explain how she won her battles but not much about how she acquired the leadership skills so necessary to her victories. Our intent is to drill deeper into not just what happened and why, but how it happened at the personal level as it relates to Joan.

Although Joan's military exploits captivated European attention and represent the core characteristics of who she was, the origins of her leadership talents remain to this day the most critical gap in the many studies of her history. Often debated are the causes and effects of her battlefield exploits; left unexplored are the causes and effects of her personal interactions on those battlefields, again the core supports for everything else in her history. Yet many characteristics of basic human nature,[4] and therefore of soldier behavior, possess common traits; traits long recognized in military history. By example, combat commanders consistently agree with General Gavin, a commander of American airborne troops during WW II, when he states, "as one always finds, in the last analysis, human nature is still the most important factor to be dealt with in war."[5] John Keegan in his book *The Face of Battle*, states, "What battles have in common is human: the behaviour of men struggling to reconcile their instinct for self-preservation, their sense of honour and the achievement of some aim over which other men are ready to kill them."[6] General Patton in writing to his son further explains, "What you must know [to be a good leader] is how man reacts. Weapons change, but man, who uses them, changes not at all."[7] In other words, leadership is derived from a set of universal norms, whether on the battlefield or in the boardroom, because humans react instinctively based on a consistent set of characteristic behaviors. Only the environment and technology change. As identified by Will and Ariel Durant in their book *The Lessons of History*, "Each *instinct* generates habits and is accompanied by *feelings*. Their totality is the nature of man."[8]

---

[3]Richey, 89-90.

[4]Particularly attractive is Michael Monahan's (*My Jeanne d'Arc*, 53) statement, "The story of Jeanne d'Arc is the most interesting in the world, because it involves all the elements of human nature, presenting them in vivid contrast."

[5]Gavin, 95.

[6]Keegan, *The Face of Battle*, 303.

[7]Puryear, 367.

[8]Durant, *The Lessons of History*, 32.

Thus history becomes a practical judgment of our elementary human experience, a normalizing of the common condition, a journey of utilitarian understanding, without proofs but with comparatives. Quoting the esteemed Carl von Clausewitz who wrote *On War*, probably the most studied and respected book on military theory yet written, "Historical examples clarify everything and also provide the best kind of proof in the empirical sciences. This is particularly true of the art of war."[9] As we progress through Joan of Arc's life using our guideposts of reason, intellect, emotion, and faith, we will also call on past military leaders to judge Joan's example, which should be one of the more enlightening aspects of our study. So let us begin at the beginning.

We defined leadership by using U.S. Army doctrine and recognizing that "Leaders operate on *instinct* that has evolved from what they have seen." Biologically, instincts are unlearned actions or reactions, but in this context they are learned through observation. What did Joan of Arc see while growing up? Where did her early instincts come from?

"Leaders operate on instinct that has evolved from what they have seen."

We would, of course, expect the strongest influences to come from her parents and family. Here Michelet is in error when he claims that Joan was "the third daughter of a peasant"[10] In fact, Joan grew up in a family of seven, with three brothers, two older, and one sister who died early.[11] Her father, Jacques, neither rich nor poor, though variously described as both by biographers, acquired some influence and an occasional political office within his small village. More importantly, he was an ardent French loyalist. A strict and careful man, he diligently protected his young daughter, which was standard practice in the Middle Ages, especially after a girl reached the age of twelve.[12]

Joan testified that Jacques, age fifty-four[13] at the time of her departure, nearly lost his senses when she ran off to join the army.[14] Curiously, two of

---

[9]Clausewitz, 170.

[10]Michelet, 7.

[11]See Warner, *Joan of Arc,* 301n, and De Vries, 36.

[12]See Charles De La Ronciere, "Tuscan Notables on the Eve of the Renaissance," in *A History Of Private Life: Revelations of the Medieval World, Vol. II.* Aries, Philip and Georges Duby, eds.; Arthur Goldhammer, trans., 285.

[13]Sackville-West, 25, Harrison, *Joan of Arc*, 21.

[14]Murray, 65.

Joan's brothers ran off shortly afterward to join with her. Likely Jacques sent them. Jacques is often chauvinized by biographers because of their modern social prejudices,[15] but is rarely characterized in the context of Joan's acquired male skills. Jacques was Joan's primary male role *model* and her accepted masculine traits as perceived by soldiers had to come from him, if not to a lesser degree from her brothers. The parish priest, Father Fronté, whom Joan must have witnessed often because of her frequent church attendance, may have also exampled a strong masculine role model. An eyewitness later wrote to the duke of Milan, "She has a virile bearing."[16] Now what "virile" meant in medieval France is hard to judge, but we might suspect she was physically athletic, resolute in her comportment, and confident around male authority—men seeing in her something of themselves, which may have been a necessary attribute since all medieval combat soldiers were exclusively male.

Although Joan lived in a very male dominated home and could be expected to acquire masculine survival talents, she remained nevertheless feminine. Her mother, Isabelle, age fifty-two[17] when Joan left home, strongly influenced her daughter's childhood and a close bond naturally developed between the two in a mutual defense against four males. Isabelle, along with the village priest and presumably Joan's voices, taught Joan her faith,[18] which is by far Joan's strongest personal characteristic, as well as her uncompromising female modesty, which suggests some early and strong sex education. By the age of twelve or thirteen, Joan knew about virginity and vowed to preserve hers.[19] As a measure of Isabelle's will, it was she who so late in life petitioned the powerful Catholic church to convene its nullification trial to restore Joan's innocence.[20] Joan's personal strength of conviction, which soldiers, priests, and kings would soon experience, likely came from her mother.

Certainly these early years of parenting were fundamental to Joan's leadership development. There were few leaders for Joan to *model* within her small village of 100 people, so her parents, and most likely the village

---

[15]See especially Oliphant, 309-10.

[16]Lynch, 107; Paine, I, *Joan of Arc*, 234,357n; Lamond, 89; Monahan, 97; Sackville-West, 11; Pernoud, *Joan of Arc: By Herself and Her Witnesses*, 98; Sermoise, 44.

[17]Harrison, *Joan of Arc*, 21.

[18]Murray, 6; Adams, 4.

[19]Murray, 63.

[20]Parr, II, 277-78; Bray, 326-27; Sackville-West, 28; Lightbody, 124,126; Marot, 26; Richey, 90; Goldstone, 241.

priest, and their attributes of character, and maybe the collective culture of the village character as well, must have exercised an inordinate influence, which Joan apparently assimilated—ergo, her "natural instincts." As the philosopher Nietzsche believed, "*instinct* is the most intelligent of all kinds of *intelligence* which have hitherto been discovered."[21] Joan was richly endowed with an extraordinary set of leadership talents, but there was something very specific contributing to these instincts that has rested for centuries surreptitiously hidden deep within the testimony taken from the Domremy villagers concerning Joan's childhood. Joan's village life was so unspoiled as to seem untrue; quoting from the nullification trial record:

"Instinct is the most intelligent of all kinds of intelligence which have hitherto been discovered."

> [*One widow testified*] She was baptized at the church of St. Remy.... she was a chaste maiden, and of modest habits. She frequented with great devotion churches and holy places....I do not think there was anyone better than she in our two villages....[Joan] had never disobeyed anyone or worked for anyone but her father.[22]
>
> [*Her godfather testified*] She was so good that all the village of Domremy loved her....it pleased her to go often to the hermitage of the Blessed Marie of Bermont....she was there when her parents thought her with the plough or in the fields.[23]
>
> [*Another widow testified*] She was no dancer and, sometimes, when the others were singing and dancing, she went to prayer. [Joan] was fond of work, spinning, looking after the house, and when necessary taking her turn at minding her father's cattle.[24]
>
> [*The church warden testified*] When I forgot to ring for service [Joan] scolded me, saying I had done wrong, and she promised to give me some of the wool of her flock if I would ring more diligently.[25]

---

[21]Nietzsche, Frederick, "Thoughts Out of Season: The Use and Abuse of History," in Durant, *The Story of Philosophy: The Lives and Opinions of Great Philosophers of the Western World*, 316.
[22]Murray, 215-16.
[23]Ibid., 213-14.
[24]Ibid., 217.
[25]Ibid., 218.

> [*One of her childhood friends testified*] Often she was bashful when others reproached her with going too devotedly to church....I loved her deeply for her goodness.[26]
>
> [*A village laborer testified*] I saw [Joan] very often in our childhood, we often followed together her father's plow....often, when we were at play, [Joan] would retire alone to 'talk with God.' I and the others laughed at her for this.[27]
>
> [*Another village laborer testified*] I was brought up with [Joan], close to her home....in my childhood I fell ill and it was she who nursed me.[28]
>
> [*A village housewife testified*] I knew her in my youth...she was my gossip, and godmother to my son Nicolas....she was not fond of playing, at which we, her companions, complained. She liked work and would spin or labor with her father.[29]

Some biographers depict Joan as a village outcast, shunned by the other children,[30] which the above testimony clearly disputes; more often Joan shunned the other children.[31] Adults respected her, it seems, and the village children desired her companionship, complaining of her independence. She followed the counsel of her voices to be good and to go often to church, but she was precocious also: she snuck away from the fields, reprimanded others, and gossiped.[32] Understandably devout, she appeared more mature than her friends or at least less frivolous. Contrary to Michelet's depiction of Joan as a homebody constantly occupied by domestic chores with her mother, she instead seems to have been more of a daddy's girl imitating his work ethic and his apparent on-duty demeanor. As testified, she never "worked for anyone but *her father*....She liked work and would spin or labor with *her father*." She minded her *father's* cattle and followed her *father's* plow. During the second day of her heresy trial, Joan herself commented, "When I was at home with *my father*, I employed myself with the ordinary

[26]Ibid., 219-20.
[27]Ibid., 220.
[28]Ibid., 221.
[29]Ibid.
[30]For example, see Taylor, *The Virgin Warrior*, 14,16-7.
[31]Lamartine, 127, agrees.
[32]This is a play on the word "gossip," understanding that in the context of this testimony it meant "friend," but which in the context of Joan's life attains the more relevant modern definition of sometimes talking too much.

cares of the house."[33] We might normally expect mention of her mother in this domestic environment.

There is a natural hardiness in this isolation of rural life, a breeding of independence in thought and action, a robust instinct for survival, a sense of personal and communal responsibility, and the sharing of values and mutual interests. It is very much a martial life without the weapons of war, which adds to the masculine influences in Joan's environment. Joan sometimes slept on the hearth "in order that the poor might be in her bed,"[34] not unlike a soldier who sleeps on the ground. Yet Joan's only real abnormality was the normalcy with which she lived her childhood, with the exception of her intense religious devotion spawned by the arrival of her voices.

But hidden within this testimony is a very important observation, "[Joan] had never disobeyed anyone." Here is a remarkably crisp statement revealing yet another vital clue bearing on Joan's leadership talent. How did Joan come by her ability to influence others so early and so effectively? Besides modeling adult examples and acquiring a strong work ethic,[35] part of the answer may be in her willing obedience to authority, as exampled by her childhood, since it was observed early in human history by the Greek philosopher Aristotle that "willingness to obey is the first step in learning to command."[36] Here then we have a crucial element in Joan's history that has been universally overlooked.

"Willingness to obey is the first step in learning to command."

The highly esteemed Confederate General Robert E. Lee, for example, was the first to graduate from West Point[37] without a single demerit.[38] As Barbara Tuchman points out in her biography of the WW II General "Vinegar Joe" Stilwell, "The core of the military profession is discipline

[33] Murray, 9.

[34] Ibid., 221. Testimony of Isabellette.

[35] "[Joan] was fond of work." Ibid., 217. Testimony of Jeannette.

[36] In the chapter on Aristotle by Durant, *The Story of Philosophy*, 67. Durant in *Caesar and Christ*, 72, and *The Reformation*, 912, continues this observation in commenting on the Romans and on the remarkably quick success of the Jesuits during the counter-Reformation. Nevertheless, Aristotle said it first. Also found in Antal, *7 Leadership Lessons of the American Revolution: The Founding Fathers, Liberty, and the Struggle for Independence*, 13.

[37] The United States Military Academy, West Point, New York.

[38] Korda, 95. Of course Grant, Custer, and Eisenhower were all showered with demerits at West Point, suggesting they were also precocious in nature (ibid., 92-3) although relatively well behaved during childhood; they just got caught more often than Joan.

and the essence of discipline is obedience."[39] General Patton of WW II fame confirms this critical inter-relationship when he says, "All human beings...have an innate resistance to obedience. Discipline removes this resistance."[40] To a large extent, learning the discipline of obedience is what modern military basic training is all about, training that often identifies and develops potential leaders.

Similarly, medieval knights were not made overnight. Young prospects, most often from noble families, were often taken from their mothers at age seven and sent to castles where they served as pages, beginning a contract of servitude, learning obedience. In early adolescence, usually fourteen or fifteen years of age, these pages were awarded a sword and promoted to squire, which initiated their more formal years of martial training while still remaining servants to the household, developing a reliable work ethic, another attribute demonstrated by Joan during her childhood. Typically, at age twenty-one, a squire became eligible for knighthood, but might never achieve that coveted rank unless he demonstrated the necessary attributes for reward, as determined by another knight who could commission him at any time.[41] Some squires were never knighted and became like noncommissioned officers, sergeants in our modern military.

Interestingly, 2,500 years ago, Confucius believed that "the foundation of a society is a disciplined individual in a disciplined family."[42] We might say the same about the foundation of an army. In this regard, Joan experienced a lifetime of instruction in the core values of her chosen profession.[43]

Of course besides her parents and her village, Joan's secret voices were also potent allies in her daily living and strong reinforcements compelling the intensity of her religious faith, but what did they teach her? Joan never really tells us. She testified that after she left home she frequently used them in difficult situations to provide specific advice or approval, but to what extent they intervened in her childhood remains a mystery. Be good, go to church, and get ready to save Orleans and crown the king are virtually all we know about their early teachings.[44]

---

[39]Tuchman, *Stilwell and the American Experience in China 1911-45*, 14.

[40]Axelrod, 174.

[41]See Eliot, 157; Freemantle, 105; Durant, *The Age of Faith*, 572; Durant, *The Reformation*, 235-36; Gazeau, 275-76; LaCroix and Meller, 28-50.

[42] Durant, *Our Oriental Heritage*, 669.

[43]"She is the image of heroic obedience." Fumet, 52.

[44]Murray, 10.

"Be good," of course, was likely the primary stimulus for her willingness to obey, which underlies the importance of her religious life—"go to church"—which also suggests that the parish priest must have been viewed by her mysterious voices as an acceptable religious teacher and role model. Otherwise we are left to wonder how it was that Joan prepared for her mission. Was she schooled in equestrian skills after protesting to her voices an inability to ride a horse? Did her voices teach her the art of warfare? Were leadership lessons a part of her instruction? It all seems possible, but only because when asked what her voices had taught her, Joan replied, "The greater part of what [they] taught me is already in the book in which you are writing"[45]—meaning her Rouen trial record for heresy.[46] This statement, delivered just a few days before the end of her formal questioning, leaves virtually the entire trial record with possible clues as to what Joan might have been taught by her voices.

Interesting to note, however, Joan's voices had occupied her teenage years for the same amount of time it takes to acquire a modern high school or college education, suggesting that Joan's flight from her village might best be viewed as a graduation rather than as running away from home, which idea returns her perceived misconduct to mainstream acceptance as an example for other teenagers to follow. In fact the king's secretary, Chartier, described Joan as appearing "to have studied at a university."[47] Here again we have *reason* and *intellect* complementing her *emotion* and *faith*. Thus when Joan's voices decided she was ready, they forced a diploma into her hands and sent her out into the world, which graduation certainly proved warranted based on her immediate and spectacular success. Nevertheless, no one really knows her saintly lessons, but what we are soon to find out, Joan possessed a miraculous ability to acquire quickly and employ expertly physical and mental technical skills, which may also suggest a genetic predisposition behind her athleticism and intellect.

Thus Joan's early leadership skills came first from her family in the form of a disciplined obedience and ambitious work ethic, reinforced by her voices and her faith in their commands to be good and to go to church, fortified by a similar conformity of character traits as possibly practiced by her village priest and maybe by the entire culture of her patriotic village. Joan of

---

[45]Ibid., 85.

[46]Parr, II, 130, interprets this passage as making reference to the book she swore on, meaning the Holy Gospel.

[47]"Epistola de puella" in Taylor, trans., *Joan of Arc: La Pucelle*, 110.

Vaucouleurs, France. (Image@2016 Google Earth)

Arc was an obedient, disciplined, confident, mature, intelligent, principled, industrious, creative,[48] and athletic teenage girl even before leaving home to join the French army. But not to forget, all of these character traits were encapsulated inside a bubble of faith in her faith and all the emotions that come with that belief system. As we will soon discover, Joan cried a lot.

As explained by Michelet, Joan eventually recruits her uncle as her emissary to Robert de Baudricourt, the captain of Vaucouleurs, a walled city of 3,000 people just twelve miles north of Domremy. Through him she acquires a six man military escort who take her to Charles VII. The most obvious observation about Joan's character while in Vaucouleurs is her persistent need to fulfill her mission—to crown the king as instructed by her voices. She influences her uncle into stalking the caustic and disbelieving Baudricourt who naturally belittled the idea of a peasant wench commanding an army or crowning a king; the woman was crazy and Baudricourt wanted nothing to do with her, rejecting her twice before eventually providing his support. There are several theories as to how Joan finally convinced Baudricourt to give her an escort to the king, but none recognize the primary cause and effect of her personal actions, the leadership techniques she applied to her environment while in that town.

[48]When asked at trial if she had learned any trade, she answered, "Yes, I learnt to spin and sew; in sewing and spinning I fear no woman in Rouen." Murray, 9.

After leaving Domremy and arriving in Vaucouleurs, Joan began calling herself "the Maid." Remarkably, this name originates from an ancient prophecy promulgated by none other than Merlin himself, that same mystical magician who so ably served as personal advisor to King Arthur, the Knights of the Roundtable, Guinevere, Percival, Lancelot, and all the ilk of Camelot in Arthurian legend, although many of these characters were invented by Chretien de Troyes, a twelfth century Frenchman who served as a troubadour in the romantic courts of Marie de Champagne. Champagne of course is the very province wherein lies the village of Domremy. Merlin, the mythology of the time supported, predicted that a young maid from the oakwoods of Lorraine would emerge to save her country. Domremy sits exactly on the Meuse River, directly along the border with Lorraine, a province sometimes belonging to itself, to Germany or to France. Of course Merlin's existence has never been proven,[49] but folklore has a life of its own and Celtic tradition transmitted this legend across the English Channel using the words *Nemus Canutum*, meaning hoary grove.[50] *Canutum* in French translates as *Chenu*, which evolved into the *Bois Chesnu*[51] or the oakwoods of France.[52] Oakwoods grew along the Meuse River valley and could be seen from the front door of Joan's childhood home, as she testified at trial,[53] and she often frolicked in these forests while growing up. Joan knew of the Lorraine legend and early in her adventures began suggesting that it was she who would fulfill Merlin's prophecy;[54] of course never mentioning Merlin himself since she likely never knew where this prophecy came from. Others were also reminded by Joan of another version of this prophecy. In one testimony, Joan is alleged to have said, "Do you not know the prophecy which says that France, lost by a woman, shall be saved by a maiden from the Marches of Lorraine?"[55]

---

[49]Merlin was invented by the 12th century cleric, Geoffrey of Monmouth, likely based on a British bard by the name of Myrddin who lived in 573 A.D. Geoffrey Ashe, "The Arthurian Fact," in *The Quest for Arthur's Britain*, 72; Taylor, *Joan of Arc: La Pucelle*, 19n.

[50]See Lang, 332n, Richards, 44, and Waldman, 16.

[51]Not to be confused with Bois Chenu which means "white wood." Julleville, 8n.

[52]Bray, 87; Lang, 32-3,332n; Bangs, 12,19; Richards, 44; Endore, 8; Waldman, 16; Fumet, 14.

[53]Murray, 21.

[54]Adams, 11-2.

[55]Murray, 227. (This particular version of the prophecy is attributed to Marie de Avignon. See Michelet, *Joan of Arc*, 11; Lang, 60; James, 30; Richards, 45,86; Paine, I, *Joan of Arc*, 26; Endore, 22,23; Waldman, 41; Sermoise, 39; Marot, 11,53,56; Meltzer, 91,91n; Taylor, *Joan of Arc: La Pucelle*, 19; Goldstone, 92; Harrison, *Joan of Arc*, 9; Castor, 234.)

Because Isabelle, the wife of Charles VI, had forfeited France to England by disavowing her son, Charles VII, in the Treaty of Troyes in 1420,[56] the metaphorical implications of the Merlin prophecy were even greater at the time. Eternal life in the Garden of Eden had been forfeited by Eve but regained by the Virgin Mary, and Joan wore a ring engraved with *Jhesus-Maria*.[57] Therefore, in a society trained in religious allegory,[58] the French people quickly profiled England as the serpent, the Troyes treaty as the apple, and Isabelle as the Eve. Concomitantly, this often creative religious culture also began fashioning Joan as the allegorical spiritual reincarnation of the desperately needed Virgin Mary to save the people of France from its demonic betrayal by treaty.[59] As early as 1429, the contemporary poet, Christine de Pizan, symbolized Joan as "a mother figure for her country," and of course, "A virgin, who is at the same time a mother, personifies the Virgin Mary."[60] As perplexing as Eve or Mary, and as mystical as Merlin, Joan's extraordinary boldness virtually began to shake the world around her,[61] especially within the small town of Vaucouleurs.

Joan alone adopted the Merlin prophecy and its "Maid" designation, which moniker she first established in her own mind before blatantly suggesting it to others. She likely learned of this prophecy in the same manner as everyone else, through general conversation and local entertainment. However, she gave it special relevance: her supposedly uneducated mind uniquely joined the rich fantasies of her regional folklore to the mission given to her by her secret voices, which remarkable creation she juxtaposed to save her country. To this day Joan's simple inventiveness remains unprecedented, and "the Maid," *la Pucelle*, remains the exclusive identity and unique property of Joan of Arc, as if universally copyrighted.[62] Today such

---

[56]Vale, 31.

[57]Murray, 40; Bangs, 43; Endore, 73; Smith, 207n.

[58]See Gurevich, A. J. *Categories of Medieval Culture*. G. L. Campbell, trans., 82-3.

[59]Whetnall, 79-80; Fumet, 14; Guillemin, 36; Barstow, 74; Wood, 141; Fraioli, Deborah A. *Joan of Arc: The Early Debate*, 125. See also Taylor, *Joan of Arc: La Pucelle*, 18.

[60]Christine McWebb, "Joan of Arc and Christine de Pizan: The Symbiosis of Two Warriors in the *Ditie de Jehanne*," in *Fresh Verdicts on Joan of Arc*, Wheeler and Wood, eds., 140. See also Lightbody, 96.

[61]Contemporary with Joan's life, the medieval Morosini chronicle states, "just as through a woman, through Notre Dame Saint Mary, [God] saved the human race, so through this young girl, pure and unsullied, he saved the most beautiful part of Christianity." Fraioli, 57.

[62]Taylor, *The Virgin Warrior*, 22, quoting Warner, *Joan of Arc*, 22-3, suggests that "maid" was a "stage of life" and commonly used and understood in this manner, but one would be hard pressed to find any trial or nullification testimony to that effect, or within the chronicles, or where other medieval teenage girls called themselves maids. In fact the letters written

ingenious branding is highly prized for its marketability. Madison Avenue should envy its 600 year legacy.

Of course the always determined and persistent Joan ignored Baudricourt's rejections and in a premeditated counter-offensive began spreading the strange news of "the Maid" arriving to fulfill this regional prophecy. Thus rumors spread until like a flash fire Vaucouleurs citizens flocked to see her.[63] Most who witnessed Joan walked away believing in her authenticity, both because they *wanted to believe* and because they saw nothing to discredit her, neither in factual detail nor in personal demeanor. Although biographers commonly attribute this populace response to the alleged excesses of medieval superstition or to the era's religious fanaticism,[64] this phenomenon might best be imagined by recalling the widespread public reaction caused by Orson Welles and his infamous radio play, "War of the Worlds," broadcast to America in the 1930's. This fictional drama depicted an alien invasion of earth from outerspace which panicked American listeners who had either failed to hear or to recall the play's opening introduction. The point is that people believe what they want to believe in accordance with the times.[65] In his book, *The Noble Savages: The Primitive Origins of Charisma and Its Contemporary Survival*, Bryon Wilson notes that "the basis of charisma is less the quality of the actual leader who comes to be called charismatic, than the *will to believe*" in that leader.[66] In this case, Vaucouleurs citizens—whether cult groupies or stolid religious worshippers, mystic or pragmatic, hopeless or impassioned—saw in Joan a populist inspiration for their needs. As with every generation forced to live their brief time on earth embroiled in civil war and foreign strife, French citizens desired an end to their constant dangers and discomforts, and if a pythoness divined a solution why not give her a chance?

---

about Joan routinely suggest the usage of this moniker, as applied by Joan, was unique to her. Although Warner, 23,280n, cites a few literary examples of its usage over several centuries by the more literate in the medieval population, it nevertheless does not necessarily imply a similar common usage among the illiterate masses, as remains the case to this day. As suggested by Murray, *Reason and Society in the Middle Ages*, 213-14, "the wealth of [medieval] documents derives from a small social group, and can therefore easily create illusions about how non-academic people thought," and spoke.

[63]For example, see Michelet, *Joan of Arc*, 15.

[64]For examples, see Lamartine, 128, James, 44-5,82-90, or Waldman, 26-7,29.

[65]"Why was the girl believed and followed? And the answer lies in the configuration of people's minds." Warner, *Joan of Arc*, 236.

[66]Warner, *Joan of Arc*, 33, citing Bryon Wilson, *The Noble Savages: The Primitive Origins of Charisma and Its Contemporary Survival* (London, 1975), 94-5 (Warner, 283n).

Can anyone really explain how such a perfect match of vision and public need takes place? An anonymous Native American once made the wonderful statement, "A good rain dance often depends on good timing," as was the case with Pope Urban II when he electrified the medieval world by calling for the First Crusade in 1095,[67] and such was the case with Joan of Arc when she called for crowning Charles VII in 1429. And that may be the best explanation we can offer. Joan's life carries the message that each of us, regardless of convention, can bring to the public forum a single idea that might suddenly tap the keg of our times; the results can be intoxicating. Joan tapped into that keg to satisfy the single greatest public thirst and desire of her times, which was to reestablish French identity and French autonomy by sanctifying a legitimate French king.

Important to realize, this early mythological process is fundamental to Joan's evolution as a leader of men. It becomes the first step in recognizing how she achieves her legitimate status, first as a soldier and then as a military commander. As will be discussed later and in more detail, theatrics often play an important part in a leader's success, and no small part of theatrics requires developing an audience who believes in the character being created. As suggested by John Keegan, "The leader of men in warfare can show himself to his followers only through a mask, a mask that he must make for himself,[68] but a mask made in such form as will mark him to men of his time and place as the leader they want and need."[69] Mounting the crown of Merlin atop the mask of the Virgin Mary began this progressive audience development for Joan, the mythological construction of a leader the people of medieval France wanted and needed—to fulfill a prophecy of promise, independence, and security. It was a process she staged with remarkable effect; and a process which reveals an *intellect*, or at least a natural *instinct*, more powerful than heretofore imagined or suspected by most biographers.

After all, in accordance with current U.S. Army doctrine, "Effective leadership begins with developing and maintaining a leader identity,"[70] which

---

[67]Lamb, 44-50; Durant, *The Age of Faith*, 587; Bishop, 92.

[68]"The word *persona* had signified an actor's mask; later it applied to the part played by a man in life; finally it came to mean the man himself—as if to say that we can never know a man, but only the parts he plays, *the mask or masks he wears*." Durant, Will. *Caesar and Christ: A History of Roman Civilization and of Christianity from their beginnings to A. D. 325*, p.394. See also Gurevich, 296.

[69]Keegan, *The Mask Of Command*, 11.

[70]*Army Leadership*, 3-5. ("Identity refers to one's self-concept." Ibid.)

also requires inspiring an audience—its wants and needs—to embrace that identity.[71] Joan's personal motives were an instinctive if not contemplative reflection of the desires within her village and, as it turns out, within Vaucouleurs and French society as a whole, which resulted in a perfect match between her vision and the wants and needs of her countrymen. One can easily see how this idea also applies to modern politics.

It might not be a surprise then to learn that this mythological process of identity development is not a medieval anomaly, strange and unique to Joan and her times. Although her Virgin Mary connection derived from her audience and the Merlin connection from a premeditated and artificial contrivance, these two processes are often integrated, as recently demonstrated by U.S. Army General David Petraeus. Before his scandalous demise as head of America's Central Intelligence Agency, he also "believed in myth-making."[72] While commanding American troops in Iraq and making a name for himself by pacifying the town of Mosul, Iraq, in 2003, his soldiers began calling him "King David." Petraeus responded, "I don't know where the King David thing actually came from, but you had to play that role a little bit"[73]—in this case making a premeditated response to an audience derived mythology.[74]

"Effective leadership begins with developing and maintaining a leader identity."

Cleverly, the King David analogy suggests comparisons with both the Biblical David and his defeat of the Philistines to save the city of Keilah,[75] and that same David who later embraced and lived among the Philistines.[76] There are also those who believe the Hundred Years War began because King Philip of France offered sanctuary to King David of Scotland from Edward III of England,[77] which King David analogy—another hundred years war—is not beyond the hubris of the American soldier. Knowledge of this modern complementary comparison makes it easier to accept Joan's

[71]Audience "want and need," as necessary to Keegan's "mask of command," are often overlooked in contemporary military writings.
[72]Cloud and Jaffe, 130.
[73]Ibid.
[74]Quincy, 74, discusses the similarities between the Biblical David and Joan of Arc.
[75]I Samuel 23:1-5.
[76]I Samuel 27:1-7.
[77]For example, see Whetnall, 30.

own emerging persona, as well as her audience and their contributions to her burgeoning leadership. This mythology phenomenon is more common than otherwise portrayed in Joan of Arc biographies and not uniquely derived only from medieval superstition or religious fanaticism—and somewhat less than miraculous.

Of course all leadership does not require myth-making, but it does require establishing what the leader stands for, which more often than not becomes his or her identity. When Joan finally meets up with her army and its soldiers, we will analyze in specifics how she transmits this identity message to her troops in a more practical and fundamental way, transforming soldier perceptions and gaining their acceptance.

Thus Baudricourt, the town captain, was confronted with a community of growing turbulence partially derived from local folklore, supported by religious allegory, and on parade throughout his city. As a result, Joan's reputation quickly expanded by what is now called urban legend, flowing rapidly across the countryside, reaching as far as forty miles away to the city of Nancy in the province of Lorraine. Soon messages were exchanged with the king's castle at Chinon, 350 miles away. Eventually the townspeople bought her a horse and some male clothing for her travels to the king, and Baudricourt reluctantly acquiesced by providing Joan with a sword and a small military escort. She was finally on her way...to immortality.

By now, however, even before departing Vaucouleurs, Joan had established some very important leadership credentials. Basic among these was her ability to accomplish set goals. In the military, commanders are given mission statements: "Invade the continent of Europe,"[78] or "Attack Hill 839," and then expected to devise their own plans for how best to accomplish each specific assignment. Successful strategic and tactical planning are not simple tasks and require substantial independent thought, analysis, and coordination. Similarly Joan had been given a mission statement by her voices and left to devise her own plans. "Go to Baudricourt,"[79] as they instructed her, was not a simple matter of knocking on his door. Joan had to evade her father in order to acquire freedom of movement and action, which she did by means of a cover story—to help with the delivery and care of her aunt's newborn. She had to win over her uncle to obtain

[78]Eisenhower's actual directive read, "You will enter the continent of Europe and...undertake operations aimed at the heart of Germany and the destruction of her armed forces," which simple words led to a 2.8 million man invasion of Europe. Ryan, 199; Korda, 433.
[79]Murray, 10.

logistic support—lodging, security, food, messaging, and transportation—which she did by use of the Merlin prophecy. And she had to gain access to Baudricourt, which she did by marketing herself to acquire public support.

Clearly the cover story and prophecy statements were premeditated acts. It stands to reason that making her cause known to the public resulted either from Joan perceiving a lucrative target of opportunity, of which she took advantage, or she intentionally recruited support from whatever quarter available, as with the duke of Lorraine as mentioned by Michelet, to agitate Baudricourt into submission. Joan was not a *simple* puppet having her strings pulled by mysterious voices as so frequently portrayed, nor did she wade headlong into her mission without considerable design. As George Bernard Shaw concluded, "Everything she did was thoroughly calculated."[80] Further, she had to combine family considerations with social expectations in a highly volatile military/political/religious environment. She walked a fine line between harlot and hero, witch and warrior, maid and madness, assassination and acclamation, all of which required reason and intellect as well as faith.

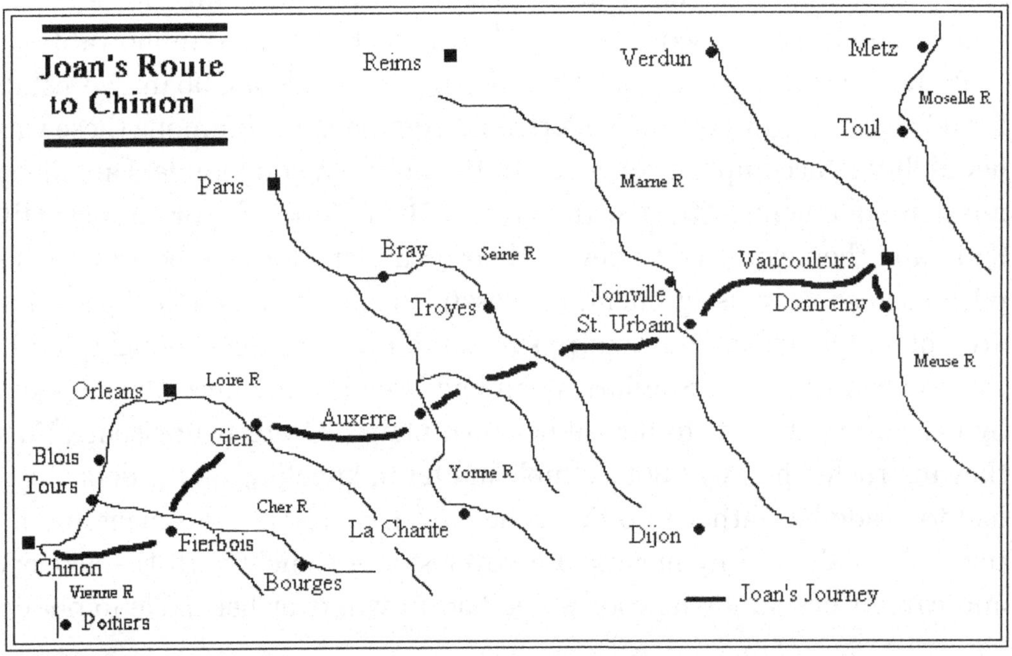

[80]Shaw, *Saint Joan*, 21; Shaw, "Saint Joan" in *Joan of Arc: Fact, Legend, and Literature*, Jewkes and Landfield, eds., 175.

Joan's subsequent eleven day journey to the king at Chinon reveals even more about Joan's leadership talents. For example, Joan possessed a marvelous ability to decipher human conduct.[81] She had a sixth sense about people, an often highly developed ability in soldiers by nature of their profession. She understood the subtleties of body language and voice inflection,[82] which suggests she was keenly observant of the people and events around her. This combination of spiritual trust and human supersensitivity was the spark of her genius. Joan's history is resplendent with cause and effect relationships associated with interpreting physical language. At the same time Joan could recognize the mannerisms of command authority in others. She gave credit to her voices but she also possessed her own observational instincts. Joan testified that when her uncle took her to the military commander of Vaucouleurs, "I recognized Robert de Baudricourt, although I had never seen him. I knew him thanks to my Voice, which made me recognize him."[83] Similarly, when Joan eventually writes to Charles VII requesting audience, she claims an ability to identify him from among all others although she had never seen him before,[84] and she does just that. Joan could recognize men of power and authority; she knew their mannerisms and carriage and could replicate it herself. Her father, and maybe the parish priest, likely conducted themselves in this exaggerated manner. We know for example that in 1423 Jacques d'Arc served as the village *doyen*, responsible to command the watch and to guard prisoners.[85] He was also a tax collector,[86] which no doubt demanded an authoritative personality. Only the mayor and sheriff were of higher rank.[87]

People who act with authority tend to be treated with all the trappings of power whether warranted or not. What is more, Joan professed direct authority from God and demonstrated a command presence to prove it. Thus her mask of command, it seems, included a complete body makeover,

---

[81]"She read character swiftly." Ince, 44.

[82]"She was keenly attuned to tiny, largely unconscious gestures and signals." Lucie-Smith, 65. See also Goldstone, 79.

[83]Murray, 11; Lang, 335n.

[84]Murray, 28; Bray, 105; Lang, 85; Winwar, 63; Leary, 26; Pernoud and Clin, 248.

[85]Lowell, 20; France, I, 25; Lang, 52; Richards, 35; Waldman, 12; Rutter, 13; Fabre, 38; Leary, 5; Raknem, 9; Wooster, 27; Pernoud and Clin, 221; Taylor, *The Virgin Warrior*, 6; Harrison, *Joan of Arc*, 30.

[86]Lowell, ibid.; France, ibid.; Buchan, 41; Bangs, 4; Richards, ibid.; Waldman, ibid.; Sackville-West, 26; Rutter, ibid.; Winwar, 22; Brown, 348; Raknem, ibid.; Warner, *Joan of Arc*, 41; Pernoud and Clin, ibid.; Taylor, ibid.; Harrison, ibid.

[87]Julleville, 5; Sackville-West, ibid.; Raknem, ibid.

which most likely came from modeling the conduct and mannerisms of her parents and maybe of the parish priest. Moreover, this idea fits perfectly again with current U.S. Army leadership doctrine: "The impression a leader makes on others contributes to his success in leading them. This impression is the sum of a leader's outward appearance, demeanor, actions, and words."[88]

Too, it is no military secret that the best men and women of power mystify themselves. John Keegan said it best in his treatise on the "Imperative of Kinship" in his book *The Mask of Command*, "Orders derive much of their force from the aura of mystery, more or less strong, with which the successful commander, more or less deliberately, surrounds himself."[89] Eisenhower, for example, was described as a "veiled man" whose mask was "secretive [and] protective of his purposes and the hidden processes of an iron logic behind them."[90] En route to Chinon, Joan likely surrounded herself in mystery. In fact she *instinctively* understood the art of mystery. After all, she had been leading a secret second life throughout her most impressionable years, exposed daily to heavenly visitations not always predictable. To protect against discovery of her secret voices, she likely normalized her otherwise mysterious conduct, disguising her assumed euphoric states with a conduct and aura of extreme piety; not that this behavior was phony, but that frequent prayer, isolation,[91] and religious ceremony gave opportunity and protection to her visitations. The children of her village teased Joan about her excessive piety,[92] but as she neared adulthood this consistency of conduct likely achieved a ritual status, natural to her character while mysterious to observers.

"The impression a leader makes on others contributes to his success in leading them. This impression is the sum of a leader's outward appearance, demeanor, actions, and words."

[88] Army Leadership, 4-1.

[89] Keegan, *The Mask of Command*, 11,315.

[90] Atkinson, *The Guns at Last Light: The War in Western Europe, 1944-1945*, 307, quoted from Larrabee, Eric. *Commander and Chief*, 419.

[91] Joan testified, "If I were in a wood, I could easily hear the Voice which came to me." Murray, 10.

[92] Murray, 220. Testimony of Jean Waterin.

More and more Baudricourt's military escort came under Joan's spell. Her eloquence of spirit "inflamed me," one of them testified.[93] Joan must have had a bit of the preacher in her and soldiers are always inspired by the powers of speech, especially when backed up by professional conduct, and Joan could inspire with words, if not Baudricourt, than most with whom she met. But most importantly, this soldier preceded his adorations with a very telling declaration, "She inspired me with such *respect*."[94] As Cervantes, most famous for his creation of Don Quixote, stated in *La Gitanilla* (The Little Gypsy Girl), "The woman who is *resolved* to be respected can make herself so even amidst an army of soldiers."[95] He must have been thinking of Joan; but again, it is a matter of *will*.

Soldiers may admire piety, speech, demeanor, and even mystery, but they respect professional conduct—actions in relation to appearance—because it is this consistency of bearing and performance that allows them to accept, trust, and then die for, if necessary, the commands given by their leaders. What is professional conduct? In a word it is self-discipline, a very sacred and revered military value, "for until a man learns to command himself it is unlikely that his command over others will prove a profitable business."[96] Stated as long ago as Proverbs 25:28, "He that hath no rule over his own spirit is like a city that is broken down, and without walls."

"For until a man learns to command himself it is unlikely that his command over others will prove a profitable business."

In the end, each soldier alone must drive himself or herself to risk life, limb, and fortune. Self-discipline is that gem of the soul that forces each soldier to engage in the survival of one to the benefit of all. It is a very human value amidst an often dehumanizing profession. As we shall see, Joan of Arc is the near personification of self-discipline.

Joan stayed in the saddle for eleven days and nights in the dead of winter, the journey conducted in late February and early March.[97] In night travel

---

[93] Ibid., 224.

[94] Ibid.

[95] *The New Grolier Webster International Dictionary of the English Language, Vol. II*, p.II, BT-102.

[96] Fuller, J.F.C. *Generalship: Its Diseases and Their Cure: A Study of the Personal Factor in Command*, 23.

[97] Caddy, 73; Gower, 28; France, I, 156; James, 60; Taylor, *Joan of Arc*, 9; Delteil, 81; Lamond, 46; Fabre, 86; Lightbody, 174; David-Darnac, 41; Cristiani, 48; Beevers, 43; Brooks, 36; Gilreath, 56.

she endured all the difficulties of reduced depth perception, unseen leafless branches scraping her skin and poking her eyes, and the tricks played on the mind that come with the setting and rising sun when the grays of nature place an enemy behind every rock and tree, even as these shifting lights and irregular shadows danced their deceptive ballets through the mists and births of nautical twilight. Joan demonstrated physical courage and endurance, an athletic ability, as well as a hearty soul. Despite whatever dangers and discomforts, she persisted by force of will, self-discipline.

Joan's ride into France compressed her basic military training, boot camp if you will, into eleven days, and as she saddled and bridled her horse for that last day's ride into Chinon she was no longer an inexperienced new recruit. She had braved enemy terrain and the night disciplines necessary to avoid hostile detection, even if the threat was minimal. She had acquired skills in horsemanship: control, direction, route selection, spurs, reins, speed, balance, rhythm, and comfort. She had not only overcome the physical elements of winter cross country travel, but had also forded at least six flooded rivers where constricted bridge sites demanded special precautions or where she had to swim or wade her mount, even while breaking through river ice. She had lived on and off the land learning the soldier's world of discomfort, how to stay mentally alert with a wet, cold, tired, sore, and hungry body. She had observed navigational skills, a bit of archery, something of the sword, and the art of range finding and distance estimation. She had learned how to travel light and fast, and how to maintain personal hygiene in the field; a soldier's primary weapon is the body and keeping it healthy amidst parasites, vermin, and predators is paramount. She had experienced the value of teamwork, cohesion, and leadership in the execution of a military mission. In the meantime, this physical ordeal increased her strength, stamina, and fitness as her previously unused muscles developed and her more common fibers toned. Not unlike modern American noncommissioned officers (NCOs) who stress "fieldcraft and physical hardening"[98] in preparing soldiers for their missions, Joan received that same training from her military escort while en route to Chinon.

Joan had also exercised her will to succeed and thereby greatly amplified what the modern military calls "mental toughness." In many ways, arriving successfully at Chinon was her graduation from basic training, and

[98]See Army Leadership, 2-2.

Joan's drill sergeants were about to release her to the next stage—advanced individual training. Ultimately, it would be said of Joan, "All the [French] captains were amazed by her courage and energy, *and her endurance*."[99]

Finally, in the larger picture, Joan also demonstrated three vital leadership talents: (1) an ability for strategic and tactical planning in relation to mission accomplishment; (2) an ability to acquire freedom of movement and action in relation to her environment; and (3) an ability to capitalize on targets of opportunity in relation to uncontrolled events. Where did these talents come from?

(1) an ability for strategic and tactical planning in relation to mission accomplishment;

(2) an ability to acquire freedom of movement and action in relation to her environment; and,

(3) an ability to capitalize on targets of opportunity in relation to uncontrolled events.

Through obedience to authority and through discipline to a work ethic, they derived from her parents and her priest, her voices and her village, all of whom developed the intellect and reason behind her faith and emotion. It was indeed a much wiser and more confident woman who would now meet with Charles VII at Chinon than the anxious little country girl who had left her family in Domremy and Baudricourt in Vaucouleurs.

Joan, as a polished new recruit, is about to ambush the French court with the surprise of its life.

[99] Lang, 154; Richards, 161.

Chapter Three

HER LEADERSHIP IN PREPARATION

We began by developing Joan's leader identity, the portrait of how she saw herself and how others saw her, and how these perceptions joined to complement each other. We examined her early childhood and the character influences from her parents, her priest, her village, and her voices; these were the foundational influences forming her leadership talents. We concluded that Joan of Arc was an obedient, disciplined, confident, mature, intelligent, principled, industrious, creative, and athletic teenage girl even before she left home to join the French army. We recognized, as did George Bernard Shaw, that "Everything she did was thoroughly calculated." We discussed her ability to recognize and replicate authority, her piety, speech, demeanor, and even her mystery, recognizing that she gained respect from her military escort through what today is called professionalism, an effective and efficient application of her leadership talents through persistent self-discipline to accomplish an assigned mission. Finally, we summarized her brief military training and the three comprehensive leadership traits she demonstrated prior to arriving at the king's castle. We can now continue by witnessing her leadership in preparation.

Joan was not immediately accepted into the king's presence. After arriving at Chinon, she was both physically examined for virginity and mentally examined by the Catholic clergy. After two days she was allowed admittance to a gathering of 300 people, one of whom was the king himself. The idea, as related by Michelet, that the king hid himself within this crowd and that Joan found him out is most likely a myth. However, it may well be that she knew him at first sight. Her instincts, her marvelous ability to decipher human conduct, likely recognized the carriage of

Charles' person, the deference others gave him, the personal space he likely commanded, and the quick glances his subjects would have secreted to gauge his royal reaction. Joan always trusted in her observational *instincts.*

But Joan was on stage, and stage acting, wearing a mask or sometimes modeling others, plays no small part in leadership in whatever profession, but is particularly acute in the business of war and politics.[1] As quoted from various biographical historians about a variety of great leaders: throughout his reign Alexander the Great "was an actor of the most consummate theatrical skill;"[2] Richard the Lionheart had a very "deliberate sense of theatre;"[3] Napoleon "felt that showmanship was a necessary prop of rule;"[4] Eisenhower wrote of Patton, "A shrewd soldier who believes in showmanship to such an extent that he is almost flamboyant."[5] Eisenhower himself admitted to having "studied dramatics for seven years under MacArthur."[6] As an aside, Eisenhower had played the role of Launcelot Gobbo, Shylock's servant, in Shakespeare's play *The Merchant of Venice*, and General Ulysses Grant had played the role of Desdemona in Shakespeare's *Othello.*[7] As William Manchester so astutely observed concerning Douglas MacArthur, "Like King David, Alexander, and Joan of Arc—like virtually all of history's immortal commanders—he was always performing."[8] We should also not forget General Petraeus in his role as King David.

To a certain extent, as with every leader, Joan too was a theater artist who could bask in the limelight upon taking the stage. After all, like Alexander who encouraged the legend that he was the son of Zeus,[9] or Caesar who traced his ancestry to the daughter of Jupiter,[10] or MacArthur whom the Japanese looked upon as "the second Jesus Christ,"[11] Joan of Arc often described herself, through her voices, as the "Daughter of God."[12] Acting of course requires getting inside the mind of an imagined character

[1]See Keegan, *The Mask of Command*, 47-54.
[2]Ibid., 90.
[3]Bridge, 130.
[4]Durant, *The Age of Napoleon*, 242.
[5]Bradley and Blair, 139; Korda, 403n. See also Axelrod, 40.
[6]Korda, 194.
[7]Ibid, 81.
[8]Manchester, *American Caesar: Douglas MacArthur 1880-1964*, 9. Books about WW II are filled with statements like: "good leaders were often good actors, able to convince their men if not themselves that they would somehow prevail." Kershaw, 59.
[9]Durant, *The Life of Greece*, 548; Keegan, *The Mask of Command*, 53.
[10]Durant, *Caesar and Christ*, 167.
[11]Manchester, 474.
[12]Murray, 64.

and transmitting that mind and character by creating a want and need in the audience for the role being portrayed. Not unlike St. Paul in First Corinthians whose goal was to be a Jew to all Jews, and unlawful to those without law, and weak to those who were weak, the idea is to be, as much as possible, "all things to all men;"[13] possibly an impossible task for anyone but a saint.

More importantly, even at her young and inexperienced age, Joan of Arc understood the value of her theatrics. Charles mistakenly put her on stage and she performed brilliantly, leaving behind in this second act a mystery as perplexing and universal as the question of Hamlet's sanity in Shakespeare.[14] Allegedly Joan told Charles VII something in private so sensitive and so secret that it has never been discovered, but which was so convincing as to inspire Charles to have her further examined. No doubt Charles must have been impressed with Joan's skillful maneuvering in achieving a job interview, and with her professed loyalty, command presence, and confident spirit of justification.

Most importantly, in accordance with current U.S. Army leadership doctrine, "Influence is the essential element of leadership. Influence refers to how people create and relay their messages, behaviors, and attitudes to affect the intentions, beliefs, behaviors, and attitudes of another person or group of people."[15] Clearly, Joan was influencing those around her, affecting their "intentions, beliefs, behaviors, and attitudes"—to include those who dealt with her in Vaucouleurs, those who rode with her to Chinon, those who examined her physically and mentally—and now, the king himself, Charles VII.

"Influence is the essential element of leadership. Influence refers to how people create and relay their messages, behaviors, and attitudes to affect the intentions, beliefs, behaviors, and attitudes of another person or group of people."

The king decided to take Joan to the city of Poitiers for a more vigorous examination by his most esteemed clerics. Not surprisingly, she passed their tests, her *intellect* and *reason* again dominating the contest. More importantly, her appearances in Chinon and Poitiers again infected the

[13] 1 Corinthians 9:20-22
[14] King David also feigned insanity. See I Samuel 21:10-15.
[15] Army Leadership, 6-2. See also Antal, 33.

townspeople with an insatiable curiosity about who this Joan of Arc was. Again rumors flourished, and again the populace became excited by the purpose of her professed mission. Mark Twain described it this way, "she had the soul of the multitudes, and her opinions were made of the opinions of all."[16] We are reminded of St. Paul and being all things to all men.

Because of Joan's remarkable history, it is hard to imagine the first impressions she made on others. She still dressed in men's clothing, all black,[17] which seems harmless enough in our own era, but fits perfectly that rebellious nature so inherent in adolescence. In the last half of the twentieth century we have seen teenagers transition from long hair, miniskirts, and beads, to grunge clothing and colorfully dyed hair, to shaved heads, tattoos, and body piercings. Each rebellion provokes shock, agitation, and attention, but none of these modern teenage revolts rivals the impact of Joan's male attire on the accepted and revered customs of the fifteenth century. Everywhere elders were stunned by what they saw. From her arrival at Chinon to the day of her death, Joan's dress and appearance provoked extreme controversy. The Bible expressly forbids women to wear garments of the opposite sex: Deuteronomy 22:5 states, "The woman shall not wear that which pertaineth unto a man, neither shall a man put on a woman's garment: for all that do so are abomination unto the LORD thy God."[18]

Nevertheless, Joan possessed a brilliance for symbolism and marketing rarely rivaled in history, although many of history's most memorable military leaders have capitalized on arrogant displays of flamboyant attire to attract the loyalties of youth and thereby win battles, wars, and nations. After all, who wants a leader with an inferiority complex?[19] Napoleon's bicorne (two-cornered) hat, Custer's long-yellow hair, MacArthur's "gold-embroidered cap of a Philippine field marshal"[20] along with his corncob pipe, Patton's ivory-handled pistols, or Native American war bonnets all come to mind, but Joan's insistence on wearing masculine dress was a quantum transformation for her time and unrivaled in its audacity. As we shall see, she even trumps her black male dress by donning knightly white armor.

[16]Searle, 86.

[17]According to the clerk of La Rochelle, (James, 86), upon arrival in Chinon, Joan was seen wearing what was described as a gray cloak, but we can suspect it was either a worn and faded black or discolored by snow, frost or dust.

[18]Even in the sixties and seventies during the modern feminist movement, the initial wearing by women of masculine-styled fashions still provoked controversy.

[19]Patton believed "We must have a superiority complex." Axelrod, 58.

[20]Korda, 212.

Again, the idea of authority and leadership can be transmitted by personal style, the appearance of the leader. Contrast for example the impeccable suits and ties of Ronald Reagan with the open collars and shirt sleeves of Jimmy Carter, or the stylish cigarette holder used by Franklin Roosevelt with the top hat worn by an already tall Abraham Lincoln.

But as with Napoleon, Custer, MacArthur, and Patton, breaking the mold and over-extending the benefits of symbolism can lead to persecution and disgrace. We love our dashing buccaneers, but when their job is done we tend to bury them amidst the tombstones of our ever changing cultural norms, forcing upon succeeding generations the problems of resurrecting their often unprecedented achievements to proffer more reflective honors. Sometimes this burial is contemporary with the times, as with Joan of Arc. Nevertheless, the singular symbol of Joan's life was her indifference to sexual identity: she was of the spirit, not the flesh; favored the soul, not the sex.[21]

Thus, during her Church interrogations in Poitiers, the clerics too became farmers of legend and saw in Joan a bountiful harvest for their coffers. It was not by accident that during Joan's time she was viewed as a possible saint. While at Poitiers these storytellers took hold and myths of grand proportions began to grace the daily pulpits.[22] As Cervantes explains through Don Quixote, "there is a crew of enchanters always among us who change and alter our deeds, and transform them according to their pleasure and their desire either to favor us or to injure us."[23] How true.

According to a letter written to the duke of Milan by one of Charles VII's counselors, Domremy roosters had crowed for two hours to signal Joan's birth.[24] During her childhood, according to the *Journal of Paris*,[25] birds and animals fed from her hands and gathered to hear her voice,[26] as had

[21]"She sees herself as a soldier and dresses accordingly. In this she is ahead of 'even' our time; in her radical approach to achieving her ends, she does not privilege gender." Meltzer, 203.

[22]"We must not be too hasty to condemn those pious frauds without which the Maid could not have worked her miracles." France, I, 180.

[23]Cervantes, 263.

[24]Perceval de Boulainvilliers. Bray, 88; Lang, 29,103; Bangs, 25; James, 31; Taylor, *Joan of Arc*, 1; Paine, I, *Joan of Arc*, 111n (see the entire letter in Paine, I, *Joan of Arc*, 352-58n); Delteil, 4; Endore, 3; Winwar, 23; Stolpe, 133; Lightbody, 56; Pernoud, *Joan of Arc*, 98; Gies, *Joan of Arc*, 10; Marot, 29; Wooster, 28; Pernoud and Clin, 55; Sullivan, 18; Taylor, *The Virgin Warrior*, 1.

[25]Lightbody, 77.

[26]Shirley, *A Parisian Journal 1405-1449.* Translated from the anonymous *Journal d'un Bourgeois de Paris*, 233-34; and in a letter from the Italian, Borosini; the "Bourgeois of Paris" in *Joan of Arc: La Pucelle*, Taylor, trans., 229; Michelet, *Joan of Arc*, 13; Parr, I, 16; Bray, 90; Adams, xi; France, I, 212,467; Lang, 34,103; Bangs, 24; James, ibid.; Taylor, *Joan of Arc*, 6; Rousseil, 10; Paine, I, *Joan*

been the case with the thirteenth century St. Francis of Assisi.[27] Marie de Avignon was quoted as predicting Joan's arrival with visions of a woman dressed in armor.[28] Merlin's prophecy was expanded by the contemporary medieval poet Christine de Pizan.[29] The magic of Alexander the Great's sword was added to Joan's legend.[30] And thunder and lightning presaged her every arrival.[31]

Joan of Arc was not always the master of her own fate, which may suggest divine intervention, but she did set in motion a product of inestimable market value as determined by the public. As with hula-hoops, cabbage-patch dolls, Pokemon cards, and Harry Potter books in our day, Joan became a fad, although unlike most commercial bonanzas she exceeded expectations: her legacy flourishes even to this day. Imagine the force of her impact almost six hundred years ago. For example, "The magistrates of Ratisbon paid twenty-four pfennigs in 1429 just for the privilege of looking at a picture showing how the Pucelle had fought in France."[32] In today's world there would have been a flurry of hats, T-shirts, dolls, figurines, stickers, books, commercials, etc., all gleaning massive amounts of capital off her popularity. Flamboyance has a commercial as well as a promotional value and Joan may have realized that fact even before our own twenty-first century—she is about to establish the most remarkable collection of regalia and symbols since the serpent, apple, and fig

"She had a genius for self-promotion unrivaled in history."

of Arc, 17,111n; Paine, *The Girl in White Armor,* 10; Endore, 11; Lightbody, 77; Pernoud, *Joan of Arc: By Herself and Her Withesses,* 101; Smith, 198n; Warner, *Joan of Arc,* 236,324n; Barstow, 53; Pernoud and Clin, 55; Sullivan, 1; Harrison, *Joan of Arc,* 28.

[27]Murray, 269-70. Testimony of Jean Barbin. Lowell, 68; Lang, 33; Bangs, 19; Taylor, *Joan of Arc,* 49; Wilmot-Buxton, 62; Endore, 79; Pernoud, *The Retrial of Joan of Arc,* 115; Vale, 50. Joan needed a John the Baptist to presage and authenticate her arrival?

[28]Murray, 269-70; Parr, I, 67; Lowell, 68; Haggard, 250; James, 30; Denis, 27; Sackville-West, 67; Fabre, 111; Stolpe, 16-7; Lucie-Smith, 75; Searle, 63; Warner, *Joan of Arc,* 24; Gies, *Joan of Arc,* 31; Leary, 30; Barstow, 64; Wood, 139; Pernoud and Clin, 30; Sullivan, 22; Meltzer, 91n; Russell, 11; Taylor, *The Virgin Warrior,* 20; Taylor, *Joan of Arc: La Pucelle,* 18; Goldstone, 92; Castor, 238.

[29]See Anne D. Lutkus and Julia M. Walker, "Christine De Pizan's Pro-Joan Propaganda" in *Fresh Verdicts on Joan of Arc,* Wheeler and Wood, eds., 153.

[30]See France, I, 226,475; D'Orliac, 126; and De Vries, 52. This extended heritage likely came from the Church because ecclesiastics were not fond of Martel: he had brazenly confiscated Church property to finance his war against the Saracens. Fuller, *Decisive Battles,* 348; Ashley, 63; Freemantle, 18.

[31]France, I, 470; See also Southey, 38,75n.

[32]Lowell, 197; Sackville-West, 1-2.

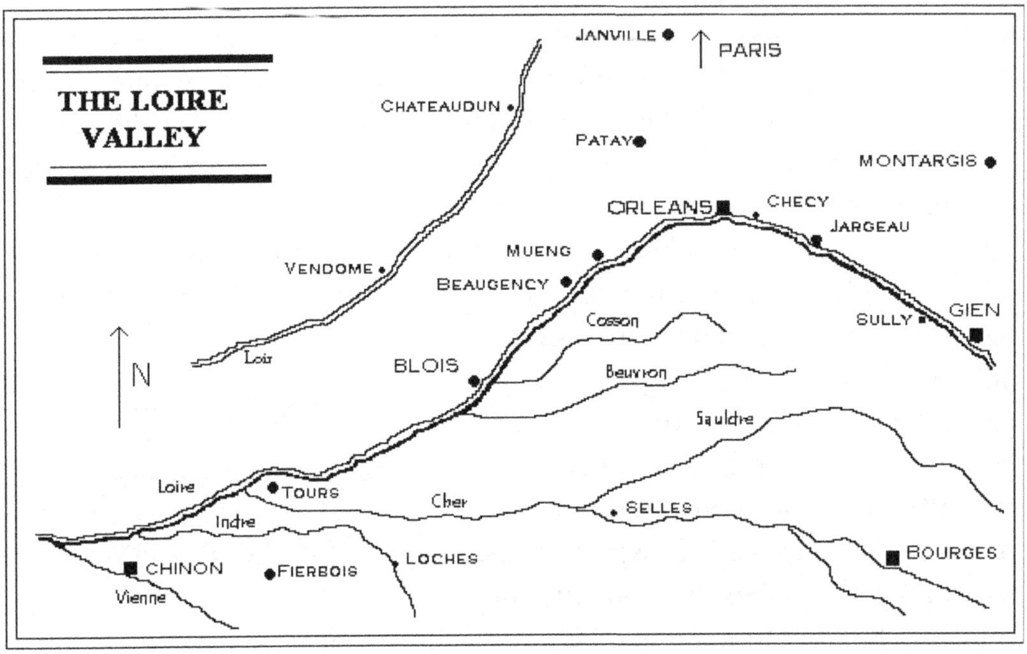

leaf in the Garden of Eden. As Francis Leary tells us in his book about Joan, "She had a genius for self-promotion unrivaled in history." Again, some very exalted praise. Little wonder the Church council at Poitiers approved her mission. It was money in the bank.

Yet this phenomenon again is not unique to Joan or exclusive to medieval history. Without having won a single battle in World War II, having barely escaped to Australia by avoiding the Japanese blockade of the Philippine islands, General Douglas MacArthur nevertheless set off a frenzy of idolatry in America. Everywhere babies, animals, food dishes, clothing items, bridges, dams, parks, streets, and even dances were named after him. MacArthur buttons, pennants, and photographs saturated the country.[33] His escape from the Philippines and subsequent declaration that "I shall return," tapped into the keg of his times and intoxicated the public. MacArthur became the leader the people wanted and needed, and their exuberance again demonstrates just how this phenomenon takes place.

After her Poitiers examination, Joan was taken to the city of Tours where she designed a gold fringed flag for herself and a banner for her following. The flag and banner were Joan's tattoos, the very essence of her universal nature as an adolescent. Modern commercial businesses call them logos.

---

[33]Manchester, *American Caesar*, 310-11.

Joan knew how to draw attention and could have done so even in our modern age. Wearing all black on a black horse, a woman dressed and styled as a man, with a newly sharpened and polished saber received from the clerics of the church at Fierbois—a church she had passed through on her way to Chinon—with a red, gold or leather sheath, gifts received from several different sources, and with her fringed gold and white flag of Christ waving for all to see, she must have made quite a sight at Tours even before her costuming was complete.

Joan was then fitted for a suit of white plated armor as if a knight of the realm. The entire ensemble, like a Madison Avenue creation, made Joan one of the most striking images ever to ride this earth: a teenage girl, adorned in white armor, astride a silky black charger, with belted sword, a fringed banner snapping boldly in the breeze with its shaft stuck firmly in

"Entrance of Jeanne d'Arc at Orleans"

by Jean-Jacques Scherrer.

her stirrup, ready for war against the greatest military adversary of her time. Add to this imagery the usual cape of red, green or gold she was known to wear and the artistry is complete.[34] Joan invented the perfect and eternal icon for female warriors, embracing all the elements of poetry, music, art, politics, economics, religion, psychology, sociology, judiciary, and military without ever having done a thing to prove herself, except *to act boldly and with conviction.* Most importantly, however, she had been uniformed and equipped by her government, signifying a soldier's official enlistment, if not an officer's commission. Like Richard the Lionheart who also took up his Crusader's oath and cross at Tours,[35] Joan of Arc took up hers.

While at Tours Joan dictated a letter—she could not read or write—which was delivered to the English commanders telling them to abandon the blockade of Orleans or die. Here Joan examples another leadership instinct. Sun Tzu, a Chinese military theorist, believed that battles were often won before any actual physical engagement, not unlike how an election campaign for political office wins the contest before the voting begins. Sun Tzu "realized that an indispensable *preliminary* to battle was to attack the mind of the enemy,"[36] an attack which began with this first letter from Joan to the English commanders. Like the daring picador on horseback who prepares El Toro for his demise in the bullring, even before arriving at Orleans Joan was at war, preparing the bull for his death.[37] In the modern military this is part of what is called "preparation of the battlefield," and here we have another example of Joan's classic military *instincts.* To the extent possible, Joan always influenced the environment around her as well as the environment ahead of her.

However, it must also be acknowledged that in her letter to the English Joan significantly expanded her mission statement. She declared that the king "will enter Paris with a good company." Moreover, she intended to drive all of the English "out of the whole of France."[38] Joan's original goals of relieving Orleans and crowning the king at Reims now included capturing the capital of Paris and conquering half the continent. Early and easy success sometimes breeds expanded ambitions. Paris and all of France

---

[34] James, 207.
[35] Bridge, 92.
[36] *Sun Tzu*, Griffin, trans., 41
[37] Patton believed "Wars are lost in the mind before they are lost on the ground." Axelrod, 55.
[38] See Murray, 36-8.

were not goals derived from her voices and, as will be seen, they abandon her during the Paris period of her life. Mission creep, constantly expanding your objectives, is a serious impediment to effective leadership. Successful leaders are constantly aware of this phenomenon and learn to avoid its emergence because it often detracts from the higher priority mission, unnecessarily dispersing resources. Usually the elephant has to be eaten one bite at a time.

"Mission creep is a serious impediment to effective leadership."

From Tours Joan traveled to the town of Blois. In American military parlance, she entered the assembly or staging area of the French army where a convoy of over 600 wagons was being loaded with supplies for Orleans. Blois was her first contact with the French army and its professional officers and soldiers. Only two months had past since Joan departed Vaucouleurs and in this short time any neighbor from Domremy who had previously known her would have hardly recognized her appearance or demeanor. She had emerged from her rural womb to breathe life into the angels of her spirit, revealing the complex design of her more urban creation.

Assigned two aides, two pages, two messengers, a treasurer, and a priest, she entered Blois with her banner held high, her entourage in tow, and her reputation embellished. Still, she was little more than a modern squad

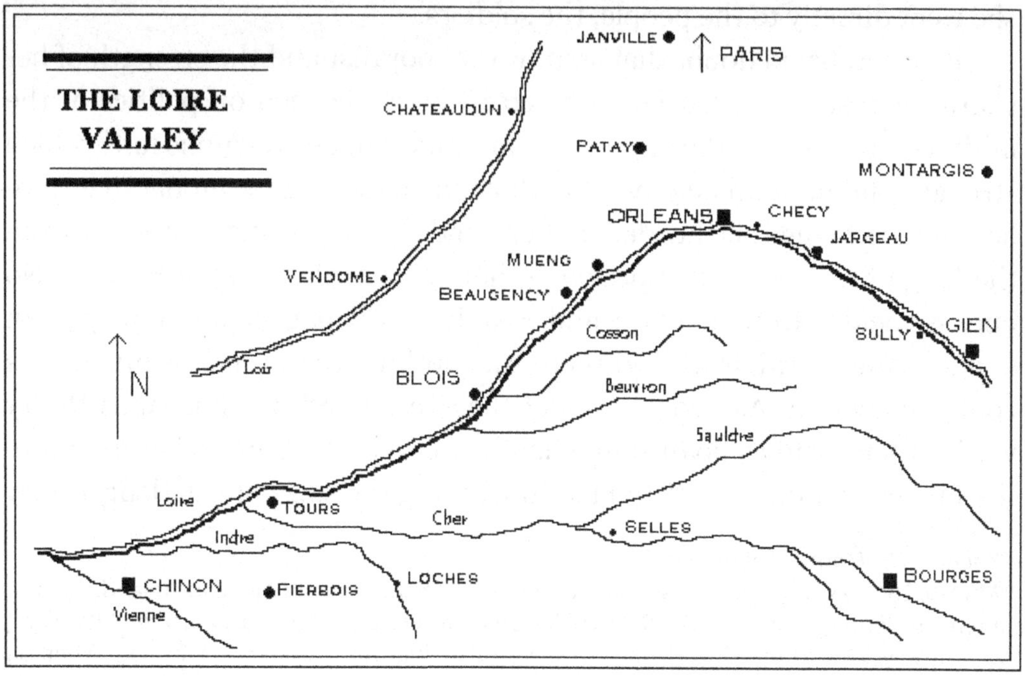

leader who had accomplished nothing more than talk her way into the hearts of a small segment of the French people and thereby win a job: no small achievement but certainly not proof she could win a battle much less a war by leading an army.

There are no direct accounts as to what transpired when Joan first met with the French commanders, but we can assume from her previous encounters, as with the likes of Baudricourt, she was not readily accepted. Joan had only one thing going for her and that was the aura of the king's seeming support. Joan was a commissioned leader with an appointment from the highest authority in the land and the king's officers had no choice but to abide by those orders or incite insurrection. To their credit they put up with Joan—out of loyalty to the French monarchy and not because of any respect for Joan's royal stamp of approval. Likely French officers reacted by wishing someone had sent her home for a good spanking.

Remarkably, Joan had been designated by Charles VII as the *Chef de Guerre*, Chief of War, but what did that title mean? No one knew and there was no job description. Whatever authority or control Joan possessed she would have to carve out for herself. As will be seen, she is immediately excluded from all military planning, which makes sense from the viewpoint of embattled professional military officers. There would be no blind faith converts to Joan's cause within the commanders at Blois. Given these circumstances, what was Joan to do? She did exactly what we should expect; she went directly to the people, the soldiers.

There can be no doubt that Joan was a moralist and the strength of her character resides in that fact. She could be challenged on politics or the military, but she was the supreme commander of her religious faith which strength she immediately employed. Joan possessed two tools that gave her all the leverage she needed; in her mind she had secular authority from the king of France, and supreme authority from the King of Heaven. No one else held both these cards and Joan instantly put them into play. Joan demanded a moral army and there was nothing the French commanders could do about it. As had Alexander the Great[39] and the illustrious Henry V with their armies, Joan immediately ordered all camp followers out of the French staging area.[40] In fact, in winning the siege of Harfleur, Henry

---

[39]Keegan, *The Mask of Command*, 65.

[40]Murray, 264,268,280. Testimony of de Contes, Simon Baucroix, and Alencon. Ireland, trans., II, *Journal of the Siege of Orleans*, 59; Michelet, *Joan of Arc*, 26; Evans, 141; Bartlett, 126; Parr, I, 222; Tuckey, 43; Adams, 62; O'Hagan, 52; Creasy, 216; Julleville, 42; France, I, 253-54; Whetnall,

V had threatened to break the left arms of any camp followers who dared to approach within three miles of his siege operations.[41] Gambling was also forbidden by Joan,[42] as had also been the case with the crusader army under Richard the Lionheart, whose punishment entailed whipping the naked soldier for three days.[43] Swearing was outlawed, as had also been the case within the armies of Richard the Lionheart[44] and Henry V.[45] Charles VII himself believed that cursing and blasphemy had contributed to the sufferings of France and he had issued previous decrees forbidding its use by French soldiers,[46] decrees Joan now enforced. These moral standards, of course, reflect Joan's parental and village upbringing as previously discussed, but are certainly not unique to Joan in the annals of military history,[47] nor a reason to label her a religious zealot.

Joan also commissioned another banner depicting the Crucifixion with the Virgin Mary and St. John on each side of the cross.[48] Twice a day, monks (who had likely joined the army to gain a daily square meal[49]) gathered beneath this banner to sing and to hold religious services.[50] Denied all other forms of recreation, Joan allowed soldiers to join in these celebrations only

78; Bangs, 91; Royden, 43; Stevens, 87; Lynch, 135; Monahan, 69; Paine, I, *Joan of Arc*, 131; Paine, *The Girl in White Armor*, 71; Endore, 86; Waldman, 83; Rutter, 138; Winwar, 87; Burne, 237; Leary, 41; David-Darnac, 58; Raknem, 110; Scott, 46; Warner, *Joan of Arc*, 233-34; Kelley De Vries, "A Woman as a Leader of Men: Joan of Arc's Military Career," in *Fresh Verdicts on Joan of Arc*, Wheeler and Wood, eds. 11; Trask, trans., 169; Ronald Edward Zupko, "The Many Faces of Joan," in *Joan of Arc at the University*, Tallon, ed., 10; Pernoud and Clin, 38; Gordon, 46; Nicolle, *Orleans, 1429*, 24; Gilreath, 70; Richey, 39; Spoto, 64; Taylor, *The Virgin Warrior*, 55-6,89; Goldstone, 124; Harrison, *Joan of Arc*, 127; Castor, 236. On the other hand, Patton believed, "An army that can't f—- can't fight," but his theatrics in this regard were an exception. Atkinson, *An Army at Dawn*, 44.)

41 Haggard, 134; Warner, *Sieges of the Middle Ages*, 183; Neillands, 207.

42 Murray, 272. Testimony of Marguerite Touroulde. Lamartine, 130; Madison, 206; Ferguson, 44; Leary, 74; Kelley De Vries, "A Woman as a Leader of Men: Joan of Arc's Military Career," in *Fresh Verdicts on Joan of Arc*, Wheeler and Wood, eds. 11,18n; Harrison, *Joan of Arc*, 127.

43 Reston, 147.

44 Bridge, 119,130; Billings, 108.

45 Creasy, 216; Haggard, 134; Waldman, 83; Burne, 54; Leary, 41; Trask, trans., 169; Ronald Edward Zupko, "The Many Faces of Joan," in *Joan of Arc at the University*, Tallon, ed., 10.

46 Lowell, 358; France, I, 253; Rutter, 138.

47 For example, during the Peasant War in Germany one hundred years later in 1525, "Blasphemy, swearing, gambling, excessive drinking, and the keeping of prostitutes in camp were forbidden." Koch, 235.

48 Murray, 30n,284. Testimony of Pasquerel.

49 France, I, 254-55; Lang, 113; Taylor, *Joan of Arc*, 80.

50 Murray, 284. Testimony of Pasquerel.

if they had first confessed.[51] Joan was quite the operative, leading by stick and by carrot, using all the strength of her abilities.[52] Clearly, Joan as a leader knew what she could and could not control.

Unrecognized by hundreds of biographers over hundreds of years, yet critical to understanding her military success, Joan was employing a very traditional military leadership technique still used to this day to establish legitimate command authority. American officers are often advised upon taking command to select some aspect of organizational discipline and quickly take ownership of it, thereby branding their new unit with their own personality while also establishing in the soldier's mind the new commander's control. Military examples demonstrating the effectiveness of this leadership technique go back as far as the Greek army 2,300 years ago. Upon Xenophon's election to his first command, as he wrote in *Anabasis*, he immediately convinced the democratic Helene army to burn their wagons and baggage trains and to reduce all other excess personal effects.[53] This discipline not only increased their mobility, but the immediate absence of baggage also permanently signaled Xenophon's control to every man. When George Washington took command of America's militia besieging the British at Boston, he was appalled at the conditions in their camp. He immediately ordered all "'stale and unwholesome food ...thrown away, clean bedding...given to the men, new latrines...dug and the rubbish around the camps...removed"[54]—quickly establishing new standards and demonstrating to every soldier that he was now in command. After the Allied disaster at Kasserine Pass in North Africa during WW II, Eisenhower relieved the failed American commander and replaced him with George Patton who immediately issued orders for all the men of II Corps to wear their helmets, neckties, and leggings at all times, and for every man to shave at least once daily.[55] By disciplining these changes in soldier routine every member of that unit understood instantly who was in command and

[51]Ibid.; Ireland, trans., II, *Journal of the Siege of Orleans*, 59; Parr, I, 86; France, I, 254; Lang, 113; Richards, 118; Lynch, 135; Taylor, *Joan of Arc*, 80; Paine, I, *Joan of Arc*, 131; Monahan, 68-9; Buchan, 93; Waldman, 83; Burne, 237; Wooster, 91; Kelley De Vries, "A Woman as a Leader of Men: Joan of Arc's Military Career," in *Fresh Verdicts on Joan of Arc*, Wheeler and Wood, eds. 11; Ronald Edward Zupko, "The Many Faces of Joan," in *Joan of Arc at the University*, Tallon, ed., 10; Pernoud and Clin, 37; Taylor, *The Virgin Warrior*, 56; Harrison, *Joan of Arc*, 128; Castor, 236.

[52]"religion is often intimately involved in warfare because it is invoked by leaders to justify aggression, to sustain morale, and to spur soldiers to the ultimate sacrifice." Wade, 9.

[53]Xenophon, W. H. D. Rouse, trans., 69.

[54]Harvey, 43.

[55]Hatch, 222; Puryear, 255; Axelrod, 17.

they felt that message every day. With Joan's presence in the French army, the sudden absence of camp followers, of gambling and swearing, and the requirement to confess and to attend mass, all had that same effect, immediately announcing her legitimate command authority, whether anyone recognized that fact or not.

Also unrecognized by previous biographers, soldier participation in mass religious services was a "synchronous activity," like marching, morning runs, group singing or even dancing within a group. Modern science has shown that these activities release the hormone oxytocin in the brain which strengthens the social attachment between participants who then become "more cooperative in completing tasks, and more willing to make personal sacrifices to benefit the group."[56] In other words, French soldiers were bonding with each other as a result of Joan's disciplines. Moreover, this bonding is complemented by the "mirror neurons" of the brain, discovered in 1996, which trigger a natural *instinct* to imitate others,[57] or what may be considered "the lemming effect." Thus, bonded soldiers join and follow each other when called into action, as will be proven later during the battle for Orleans.

Joan's instant impact on the character and personality of the French army can hardly be described as simple. Many of these soldiers were seasoned veterans with natural expectations. Men who risk their lives expect a bit of fun before they do so. Surely Joan must have met with some resistance,[58] but face-to-face with these soldiers, often nothing more than ruffians, she successfully prevailed—not always an easy task even for veteran commanders.

Nevertheless, like Xenophon, Washington, and Patton, Joan contradicted the martial norms of her time and invoked strict standards of disciplinary conduct. Her message was that if you are going to risk your life, you best be prepared to meet your Maker, which concept of course is classic military psychology for any man preparing for battle, as expressed by the pre-Christian military commander, Xenophon:

---

[56]Weinschenk, 25-6.

[57]Ibid., 21; Kaku, 55.

[58]"Leaders increase receptivity to messages and build support for achieving goals by interrupting the other party's focus on resistance. Refocusing on intended positive outcomes may sway negative mindsets and consume some of the energy fueling resistance." Army Leadership, 6-3.

> ...when men seek for nothing in warfare but only life at all costs, they are generally the ones to die, and that with disgrace...but when they recognize that all men must die, for this is their common lot, and strive only to die with honour, these I generally see growing to old age, and while they live, much happier.[59]

In Joan's Christian era, dying with honor meant acceptance by God.

Although all of this psychology may sound uniquely Athenian or medieval, the twentieth century Patton fought his battles believing, "It is a sin not to kill if we are serving on God's side. There is no other way to win....[60] We can take any fear if we are not fearful of death."[61] Of course every soldier fears death, but to conquer that fear is to conquer all fears and thereby increase his chances for survival. Fear is a relative *emotion* and fear of death can be sublimated. Besides, a soldier's job is not to die for his country, but rather to make the other poor dumb bastard die for his country, as the saying goes.

More significant, however, like Xenophon, Washington, and Patton, Joan was concentrating on the spirit of the soldier not on his body, and that is the important point. When General Eisenhower was assigned as the first NATO commander, he addressed a joint session of Congress. *The New York Times* noted: "It is further characteristic of General Eisenhower's point of view that he puts his reliance not on armies alone...but in things of the spirit."[62] In accepting the Japanese surrender on the battleship Missouri at the end of WW II, General Douglas MacArthur talked of "human character" and warned, "It must be of the spirit if we are to save the flesh."[63] Soldiers learn to take care of their bodies and equipment; *leaders learn to take care of the soldier's spirit*. Taking care of the spirit requires discipline.

Pope Pius II, supposedly a spiritual expert, also recognized the disciplinary aspects of Joan's leadership. In considering the possibility of a conspiracy behind the emergence of Joan, he suspected this motive: "It was thought that a virgin with a divine mission would improve *discipline*, for what man refuses to obey God?"[64] Now this was a highly original thought

[59]Xenophon, W. H. D. Rouse, trans., 64
[60]Williamson, 87.
[61]Ibid, 160.
[62]Korda, 620.
[63]Harvey, 340.
[64]Gragg, trans., 208; "Commentaries of Pope Pius II" in *Joan of Arc: La Pucelle*, Taylor, trans., 352; Julleville, 168.

in terms of categorizing Joan's role in the French army, but seems perfectly valid. As previously discussed, discipline is vital to the soldier ethos.

Again citing U.S. Army leadership doctrine, "Leaders who consistently enforce standards simultaneously instill discipline that will pay-off in critical situations."[65] In other words, when discipline to standards permeates an army, or even a civilian organization, it creates the possibility for unity of action.[66] As Antoine-Henri Jomini, another military theorist, observed, "Concert in action makes strength; order produces this concert, and discipline insures order; and without discipline and order no success is possible."[67] Michelet brilliantly recognized that "unity of action"[68] was the missing ingredient in French military tactics until the arrival of the Maid.

"Leaders who consistently enforce standards simultaneously instill discipline that will pay-off in critical situations."

Further, Joan's disciplines[69] also had a more curious impact as revealed by another French military theorist, Ardant du Picq, who reasoned that "*Discipline itself depends on moral pressure* which actuates men to advance from the sentiments of fear or pride."[70] This increasing advancement away from fear through Joan's use of moral pressures to discipline the French army would soon prove crucial in the battle for Orleans.[71] As du Picq reminds us, "In the last analysis, success in battle is a matter of morale,"[72] which proves true in almost every endeavor requiring leadership.

[65]Army Leadership, 6-6.

[66]Michelet, *Joan of* Arc, 25, also recognized this important law of warfare.

[67]Jomini, 42.

[68]Michelet, 25.

[69]Delbruck, 287, makes the sweeping statement that "the Middle Ages never considered establishing true military discipline." However, for Delbruck, true military discipline meant the ability to maneuver on the battlefield in unit formations known as battle drill, (Ibid., 288.) which chairborne analysis ignores the need for soldier discipline in conjunction with unit discipline, of which one cannot be achieved without the other. Delbruck's idea that nothing would have been gained through medieval military discipline is disproved by Joan.

[70]Du Picq, 96

[71]"What is called discipline is that when encamped [the army's] conduct is proper." Griffin, trans: *Sun Tzu, The Art of War*, 159.

[72]Du Picq,109. This fact has been proven true from as early as the battle of Plataea in 497 B.C. where "the psychological factor, loss or gain by the soldier and gain or loss of prestige by the supreme command was, as it still remains, the determining factor in war." Fuller, *Decisive Battles*, 5.

The French commanders could not object to Joan's intrusions. They had neither the moral nor the political authority to do so. Consequently, in the space of less than three days, the soul of the French army belonged to Joan and the body of that army was transformed. Business as usual was not an option and no one could evade Joan's pervasive *influence*. Soldiers were preparing to risk their lives the way she thought they should, with a cleansed soul. This energy of spirit and purpose, inherent to her personality, electrified those around her.

By all accounts the French soldiers accepted Joan's edicts; after all, many in the militia had volunteered for religious reasons based on the Church sanctioned validation of her divine authenticity as awarded at Poitiers.[73] Of course some had joined the army for economic reasons so as not to starve; others were seasoned soldiers conditioned to accept orders from whatever leader they were assigned; and some simply followed the crowd when told to do something. But this soldier acceptance of Joan's religious legitimacy magnified her military authority, permeating the French army and giving the French officers little choice but to follow. Brilliantly, Joan was usurping that one ingredient in a combat unit essential to victory—the spirit. In so doing, she influenced the minds of the French soldiers to accept the concept that God was on their side so long as they joined with God,[74] which concept she also used to instill other disciplines. Thus through moral pressure Joan was already bolstering that very aspect of an army that is the least predictable of all the many weapons brought to the battlefield—the will to fight. The results of her efforts would come to an amazing glory at a single critical instant during the battle for Orleans.

Developing a soldier's will to fight, however, is often a mysterious problem with more than one solution,[75] but in this case we find a curious correlation. Custer prayed before every battle, as many others have done before and since, but he also wrote about it to his wife, saying:

> Never have I failed to commend myself to God's keeping, asking him to forgive my past sins, and to watch over me while in danger....After having done so all anxiety for myself, here or hereafter, is dispelled. I

[73]Fraioli, 193, argues that Joan's "approbation" (approval) at Poitier's was for purely religious reasons, which is not entirely true, because it also served a vital military function.

[74]There are those who believe that religion itself "evolved as a response to warfare." Wade, 233.

[75]"Army leaders can draw on a variety of methods to influence others and can use one or more methods to fit to the specifics of any situation." Army Leadership, 6-1.

> feel that my destiny is in the hands of the Almighty. This belief, more than any other fact or reason, makes me brave and fearless as I am.[76]

Achilles prayed to Zeus before the walls of Troy.[77] Alexander the Great sacrificed daily to the Gods.[78] As Machiavelli reminds us, "the Romans inspired confidence in their men by means of religion."[79] In preparation for his famous sea battle at Trafalgar where he died, Admiral Lord Nelson of the British navy wrote a prayer committing "my life to Him who made me and may his blessing light upon my endeavours for serving my country faithfully."[80] Before his miraculous victory at the Battle of Agincourt, Henry V heard mass three times in succession[81] and all of his soldiers confessed their sins.[82] Both Patton[83] and MacArthur[84] read the Bible daily. As revealed in a combat survey of soldiers from WW II:

> An important motivation within the soul of the combat soldier was prayer. This is not to say that American soldiers were on a religious crusade in World War II. They did not fight for God or to spread any particular religion or creed....Praying gave combat soldiers a crutch to lean on in battle so that they could face the unimaginable terrors of war....Faith in God probably reassured American combat soldiers that there was indeed some sort of order or reason to the madness of war.[85]

And this is not just a western idea. Chinese leaders held a prayer vigil before attacking Pork Chop Hill during the Korean War, which was overheard by the American soldiers defending that vital piece of terrain.[86]

Maybe not for all men, but maybe for many, Joan's actions in the French staging area, which included spiritual cleansing through prayer, confessions, and attendance at mass, had that same effect, inspiring a sense of security and invulnerability that leads to courage.[87] From Plato, "we are

[76]Connell, 112.
[77]Atkinson, *The Guns at Last Light*, 466.
[78]Keegan, *The Mask of Command*, 47.
[79]Bondanella and Musa, ed. and trans., *The Portable Machiavelli*, 208,406.
[80]Harvey, 119.
[81]Keegan, *The Face of Battle*, 115.
[82]Evans, 71; Bartlett, 66.
[83]Williamson, 59.
[84]Manchester, 3. See also Puryear, 311.
[85]McManus, 273.
[86]Marshall, 81.
[87]"She required that freedom from fear provided for in confession so that no arm should be unnerved in battle by fear of death." Stevens, 82.

twice armed if we fight with faith,"[88] to Patton, "if the day of war does come again, you will find strong support in religion,"[89] prayers have permeated the battlefield. Perceptions of Joan's religious authority validated her actions and served this military purpose.[90] We should not be surprised then to find that when Joan finally calls upon her soldiers for a final attack during the decisive moment at Orleans, *The Journal of the Siege of Orleans* records that each Frenchman fought "as if he thought himself immortal."[91] Should we really be surprised that football players routinely join in a circle of prayer before engaging in their brutal game.

The metaphorical military lesson here might be that if one burns the baggage, kicks out the camp followers, cleans up the camp, puts on helmets and ties, stops the cursing, quits the gambling, and adds a little prayer, the organizational spirit can be disciplined to fight and win in battle—or in corporate competition. The idea is to find a pervasive change that is not only productive but also affects everyone in the organization, and then implement it.

But all of these tricks of the trade, so to speak, may reasonably be reduced to one simple fact: Joan was giving all of her attention to the welfare of her soldiers before embarking on her mission.[92] Soldiers—or anyone else for that matter—begin to care for their leader as soon as they recognize that he or she cares for them. Patton believed that eighty percent of a commander's job was to "arouse morale in his men."[93] Joan's personal attention to as many soldiers as possible inspired a sense of loyalty between them. And these personal loyalties become group loyalties, essential to an army. Soldiers *follow orders* either because they have a sense of duty or a sense of loyalty to their leaders, which can be defined as a sense of honor; soldiers *fight* either because they have a sense of duty or a sense of loyalty to their comrades, which can be defined as a sense of morale. The glue that binds honor and morale together is the mutual spirit of reliance that develops between soldiers and their leaders, and this spirit is what creates the will to fight. When called upon to win battles, it is not just numbers alone but

[88]Durant, *The Story of Philosophy*, 24.

[89]Puryear, 317.

[90]This reliance on religion to fortify the spirit, however, does not imply a motive for conducting wars; individual survival as opposed to collective or tribal survival often requires different means to achieve the same ends.

[91]Waldman, 128.

[92]"Leaders illustrate through their presence that they care." Army Leadership, 4-1.

[93]Axelrod, 108; Atkinson, *An Army at Dawn*, 138.

the numbers who actually step forward who make all the difference. When Joan finally makes that call, almost all of her soldiers leap forward to fight.

Thus, through her use of moral and religious disciplines Joan creates a personal bond with her men and as John Keegan notes in his book *The Face of Battle*, "the personal bond between leader and follower lies at the root of all explanations of what does and does not happen on the battlefield"[94]—or in business and politics.

Robert Harvey, in his book *Maverick Military Leaders*, concludes that "Great military leaders have often shown an almost *political* ability to charm and inspire their men rather than rely merely on handing down orders."[95] As soldier turned politician, Charles de Gaulle observed, "the statesman must concentrate all his efforts on captivating men's minds,"[96] which is also what an actor must do on stage. This seems to hold true whether a politician or a military commander: it is easy to see how they bleed into each other. Joan was so effective at captivating men's minds that we remain captivated by her even today.

On April 27, 1429,[97] the French army at Blois set out for Orleans, Joan with her priests and monks leading the way, which again signals her control and further demonstrates her emerging military leadership.[98] The banners of the Crucifixion, Annunciation, and of God flapped briskly in the spring breeze as she led what was to become her army. From within this mass of gray-brown procession, undulating beneath a heavy foam of white flags, arose the song *Veni Creator Spiritus*[99] (Come Holy Spirit), as if the very seas of Joan's soul were on the march. The rest of the supply convoy dutifully followed, waves lapping the shore of her storm. No doubt the coiled

---

[94]Keegan, *The Face of Battle*, 114.

[95]Harvey, xxv.

[96]Phillip C. Naylor, "Charles de Gaulle and Joan of Arc: A Comparison of Person and Praxis," in *Joan of Arc at the University*, Tallon, ed., 118.

[97]Various authors employ different dates, for example, April 26, (Sermoise, 71), April 27 (Waldman, 84; Sackville-West, 150) or April 28 (Parr, I, 93).

[98]By riding in front, Joan also maintained a more pristine personal appearance than would her dirt and dust covered followers, adding to her image of personal discipline.

[99]Murray, 284. Testimony of Pasquerel. Ireland, I, cxln; Parr, I, 93; Bray, 124; Tuckey, 44; Caddy, 167; Lowell, 95; Julleville, 36; France, I, 256; Lang, 113; Bangs, 92; James, 67,210; Wilmot-Buxton, 73; Stevens, 86; Richards, 118; Lynch, 138; Taylor, *Joan of Arc*, 80; Rousseil, 182; Paine, I, *Joan of Arc*, 133; Lamond, 62; Endore, 88; D'Orliac, 146; Waldman, 84; Rutter, 148; Buchan, 94; Winwar, 87; Stolpe, 109; Leary, 42; Guerin and White, 84; David-Darnac, 59; Raknem, 13; Gies, *Joan of Arc*, 61; Maxwell, 56; Pernoud and Clin, 38; De Vries, 71; Nicolle, *Orleans, 1429*, 50; Taylor, *The Virgin Warrior*, 61; Goldstone, 124; Harrison, *Joan of Arc*, 130; Castor, 104. The words for this song are in Parr, I, 96.

winds of the Virgin Mary and Merlin swirled about in their turbulent currents. It was an opening scene worthy of Shakespeare.

Joan was a master of theater and of that talent she left no doubt. She symbolically crossed what the American military calls "the line of departure" and thereby entered onto the field of conflict. She did so with great fanfare: a parade of priests, flags, soldiers, wagons, cattle, sheep, horses, and knights, all punctuated with the sparkle of lance, stirrup, armor, and steel. The glint of colorful parade in the early morning mists harmonized in a call for combat as the music of voices combined with the clumps of hooves, creaks of wheels, rustles of leather, and pings of metal. French officers likely sat chagrined at this display, hardly inspired by what looked like a farmer's market instead of a combat unit, yet begrudgingly respectful of the fact that she was present at all. Even at this moment Joan knew no allies except her voices. She remained a single, isolated phenomenon whose future was yet to be determined, and whose ultimate survival now hung precariously on the black clouds of courtly intrigue. As the convoy slowly drifted into the distance, the last line of their song echoed across the hills of France: "Send forth again Thy creating Spirit, O God, and renew the face of the earth."

Joan was no longer just *Chef de Guerre*. "The Maid" had arrived and soldiers were beginning to adopt her chosen moniker. As her history progresses, fewer and fewer would ever refer to Joan of Arc by any other name. "La Pucelle" would transcend her Christian baptism and royal designation. "The Maid," unique in all of history, would come to mean only Joan of Arc, Commander of the Armed Forces of Medieval France—and she was still without a job description.

Within the week she would surprise the English by winning her first battle.

Chapter Four

HER LEADERSHIP ACTION

Previously we observed Joan's leadership in preparing for her mission, recognizing: her theatrical nature and its influence on the "intentions, beliefs, behaviors, and attitudes" of others; her flamboyance in attracting attention to gain public awareness, acceptance, and support; and her genius for self-promotion, creating a unique corporal appearance and spiritual reputation even before showing up for her job as *Chef de Guerre* of the French army. We saw her prepare the battlefield by writing a letter challenging the minds and morale of her English opponents and how this letter contained mission creep—the unfortunate expansion of her original goals. We witnessed her ability to grab control of the army's daily routine by eliminating all camp followers, gambling, and swearing in the army, and by requiring religious confessions before attending church services, all of which established her legitimate command authority. And we explored a few tricks of the trade when it comes to leadership, recognizing that Joan gave priority to the welfare of her soldiers, creating a personal bond with her men. We can now explore how all of this pays off on the battlefield—or in the workplace.

To understand Joan's history, it is necessary at this point to spend some time discussing the strategic and tactical environment surrounding the blockade of Orleans. Suffice it to say that Michelet is generally correct in his view of the English army as a weakened combat force, especially after the Burgundian rebels withdrew their 1,500 supporting men due to a dispute with the English over an alleged surrender proposal offered by the city of Orleans. Nevertheless, several of his facts are wrong.

As with most Joan of Arc biographies, the circumstances surrounding the Orlean's blockade suffer from exaggerations and logic failures.

For example, Michelet's claim that only two or three thousand English surrounded Orleans represents the lowest strength estimate of any historian. Through independent analysis, this English force was closer to 4,500, which by coincidence is the average of all other historians combined. The French, on the other hand, likely had about 2,500 amateur militia inside the city,[1] which had a normal civilian population of about 15,000 with the possibility of an equal number of civilian refugees inside at the time. Of course the French also had close to 4,000 soldiers in their army at Blois, including approximately 500 volunteers. These numbers, 6,500 French versus 4,500 English, do not offer an overwhelming nor obvious advantage as alleged by Michelet,[2] especially since the number of professional or mercenary soldiers, approximately 3,500[3] French versus 4,500 English, favored the island invaders.

Nor were the string of English forts unable to support each other as Michelet alleges.[4] By fact, several attacks by the French in the course of the seven month blockade—against supposedly isolated outposts—were repulsed when English troops responded either with a mobile reserve or from immediately adjacent outposts

Michelet also praises the prestige of the French commanders while ignoring the equally qualified English commanders on the other side. For example, Suffolk, Scales, and Talbot were all distinguished enough to wear the prestigious Order of the Garter,[5] and Talbot was considered by the Burgundian captain, Jean Wavrin, as "the wisest and most valiant knight of the realm of England,"[6] and by Pope Pius II as "the most celebrated of the English commanders."[7]

Thus Michelet once again relies on *emotions* rather than *intellect* to deflate Joan's achievements to a more expected level where the defeat of the

[1] Lynch, 133. French forces inside Orleans fluctuated depending on time. Parr, I, has them growing from 500 (p.30) to 7,000 (p.49). Winwar, 98, and Harrison, *Joan of Arc*, 140, estimate an eventual total of 5,000. Whetnall, 86, counted 7,000, and Bray, 79, 12,000. Burne, 229, counts 2,400 French and 3,000 militia. Likely there were 2,500 Orleans city militia, with 3,500 professional and mercenary French troops joined by another 500 volunteers, making a total of 6,500 after Joan entered the city with her convoy.

[2] Michelet, 24.

[3] This number includes Clermont and his 2,000 soldiers as well as an estimate of the forces under Boussac, La Hire, and de Rais.

[4] Michelet, 24.

[5] Pernoud and Clin, 201,202,203; Pollard, 123.

[6] Paine, I, *Joan of Arc*, 218; Gies, *Joan of Arc*, 97.

[7] Gragg, trans., 204.

English is virtually inevitable, and where we can see ourselves equally capable of winning her battles if given the chance. However, there is a more realistic interpretation of the situation around Orleans before bringing Joan back into our considerations.

Let us begin with the broader strategic situation within France at the time. A constant historical mistake made by most political leaders throughout history, in this case by both the English and by the rebel Burgundians, is the belief that possession of a nation's capital leads to victory,[8] when in fact military success requires the surrender,[9] annihilation or the debilitating attrition or economic immobilization of the opposing armed forces.[10] The duke of Burgundy made exactly this mistake when he advised the English duke of Bedford that "Paris is the heart of France. The loss of that city would mean the loss of the kingdom...if [our] adversaries cannot be *sent to a distance and a tract of country round about kept free of them*, then [Paris]...will be unable to support and save itself."[11] In short, to lose Paris was to lose France, which eventually proved untrue. Nevertheless, this idea of creating a buffer zone around Paris was most likely the English military strategy in France when Joan of Arc entered Orleans.

Nonetheless, this hysterical attitude concerning capitals is common in history as best exampled by the Napoleonic war of 1812. Many Russian generals and politicians were of a similar belief that "Moscow's fall meant the collapse of Russia, the triumph of the Antichrist...and the end of the world,"[12] which again proved untrue. The fact that the English and Burgundians felt the same about Paris is one reason why the French ultimately prevailed and the English failed. Have we not seen this in our own time, as with Baghdad, Iraq?[13] Charles VII was never particularly inter-

[8]Burne, 225, makes this same mistake by believing that it was necessary to attack Orleans because it was the shortest distance to Bourges, the French capital, which he believes was the correct English objective. By more recent example, during WW II, "Churchill hammered away at the urgency of taking the German capital [Berlin]," which Eisenhower nevertheless ignored. Ryan, 253. "Victory doesn't come automatically with the capture of the enemy capital." Powell, 210. Nor does victory come automatically with the recapture of a friendly capital as learned by Wellington when he elected to retake Madrid after his brilliant success against the French at Salamanca in 1812. Harvey, 216.

[9]"To capture the enemy's army is better than to destroy it." *Sun Tzu*, Griffin, trans., 77.

[10]"In war, the subjugation of the enemy is the end, and the destruction of his fighting forces the means." Clausewitz, 526.

[11]Lang, 225; Bangs, 235; Wilmot-Buxton, 123; Fabre, 237. See also Castor, 82.

[12]Lieven, 235.

[13]The capture of Baghdad did not end the war in Iraq because the Iraqi army was not controlled and neutralized, allowing many of its members to reform as ISIS.

ested in recapturing the capital of France and spent most of his time and resources correctly going after the English militarily and the allied rebel Burgundian resistance politically.

Thus, with their occupation borders secure in the north and east, the English began a two-pronged military offensive to push the French west and south away from Paris, eventually reaching Orleans in the south in October of 1428 where they proceeded to tie up the French army to keep it from buzzing around the capital of Paris. Nonetheless, most historians believe this English operation was designed to capture Orleans and drive Charles out of France by attacking into the French interior from what would become an English base at Orleans. As a result, endless Joan of Arc biographies describe Orleans as the "gateway" to conquering all of France, and that "to lose Orleans was to lose France," thereby juxtaposing the actual military situation by replacing the strategic defense of Paris with a tactical offensive siege of Orleans as the real purpose of the English operation, which idea of course embellishes Joan's reputation as the savior of France when she saves the city of Orleans. It is all commercial hogwash. In fact, after arriving at Orleans, the English quickly crossed the Loire River to capture the Tourelles, a two tower structure that guarded the bridge leading to the southern gate into the city, proving they never needed Orleans or its bridge to cross the Loire River for an invasion of

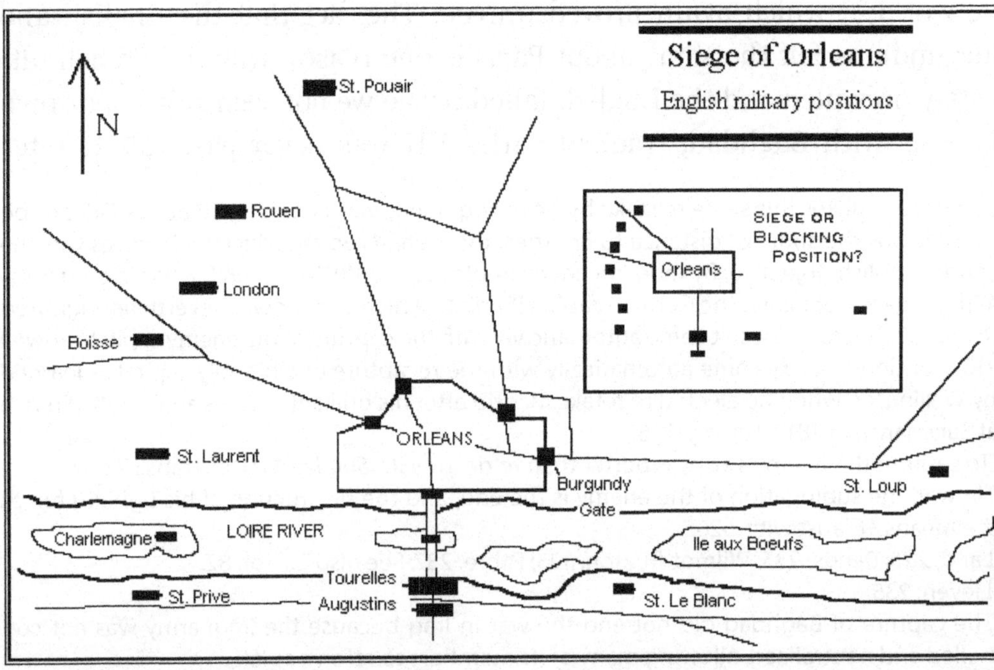

the French salient. Instead, this maneuver cut the approach on Paris and completed the English mission. The English then dispersed their forces to other towns and villages all along the Loire River for the same purpose of securing the adjacent approaches on Paris.[14] They then built the forts of St. Prive, Charlemagne, and St. Laurent, not to encircle and besiege Orleans or to block any transportation routes as some allege, but simply to secure an English supply line from the Tourelles to Paris. After all, as Frederick the Great recognized, "the first consideration of the general must be for his provisions."[15]

A simple map survey clearly shows that the remaining forts were also constructed to link these English forces with the road to Paris, using a trench-to-fort depot system, which secured this north-south supply line. *The Journal of the Siege of Orleans* specifically states that the forts were built for "conveying food one to the other."[16] This purpose also explains why the northeastern and eastern approaches to the city were left open, a situation that confuses historians who consistently assume the English objective was to encircle and take Orleans rather than to protect Paris.

Meanwhile, the fort of St. Le Blanc, constructed six months later on March 10, 1429, together with St. Prive, provided flank security for the Tourelles. St. Loup, not occupied until seven months later on April 20, 1429,[17] just nine days before Joan's arrival in Orleans, did nothing more than increase outpost security along the Loire River, further isolating Orleans by detecting and dissuading any French shipping or large-scale river crossings,[18] thereby leaving a three mile wide gap inland between the nearest river outpost and the northernmost fort on the eastern side of the city leading to Paris. This gap was consistently used by the French to re-supply the city and was never blocked by the English. The St. Loup position, however, was obviously the most isolated and vulnerable of all the English outposts.

The English dispositions at Orleans therefore formed an elongated L-shaped protection against any French attacks on Paris emanating from

[14]On November 8, part of the English army was sent to Meung and another part to Jargeau, leaving only Lord Scales and 500 English inside the Tourelles, content to occupy the two towered structure guarding the Orleans bridge. Ireland, trans, II, *Journal of the Siege of Orleans*, 7; Michelet, *Joan of Arc*, 35; Lowell, 86; France I, 130; Lang, 69; Waldman, 89; Rutter, 143.

[15]Luvaas, ed. and trans., *Frederick the Great and the Art of War*, 310.

[16]Ireland, trans., II, *Journal of the Siege of Orleans*, 13.

[17]France, I, 231; Lang, 95.

[18]Burne, 247, agrees with this assessment concerning St Loup.

within the French salient while also controlling any river traffic that might otherwise contribute to Orleans as a base for French operations[19] because of its greater capacity and shorter distance to the French capital. Nonetheless, nowhere will you see this English military strategy recognized in any Joan of Arc biography. After all, saving France is Joan's legacy and to suggest that Orleans was not the pivotal event in this rescue is historical and religious blasphemy. Although Joan does save France, it is not because she saves Orleans.

Thus, from a purely modern professional military perspective, the attack on Orleans was never intended as part of an *offensive* campaign but instead was a maneuver to establish a *defensive blocking position.*

U.S. Army doctrine defines a blocking position as "A mission assigned to a unit which requires it to stop the attacking enemy from advancing through a specified location, usually for a specified time. Units assigned this mission may have to retain specified terrain and accept decisive

"BLOCKING POSITION: A mission assigned to a unit which requires it to stop the attacking enemy from advancing through a specified location, usually for a specified time. Units assigned this mission may have to retain specified terrain and accept decisive engagement."

engagement."[20] On the other hand, if the English had intended a siege, we should expect to see the deployment of siege equipment, such as battering rams, siege towers, mobile sheds, mangonels, trebuchets, catapults, and even mining operations as was the case with Henry V at the siege of Harfleur in 1415, or the use of chains[21] strung across the river to block its use for resupplies, as was the case with Henry V at the siege of Rouen from July 1418 to January 1419. But the English ignored these siege techniques and equipment at Orleans—because it was not a siege.

---

[19]Marshall Foch in "For Joan of Arc" in *For Joan of Arc*, 5, thought the English wanted Orleans for its own base of operations, but the dispersal of their attack forces does not support this idea.

[20] *RB 101-999: Staff Officer's Handbook.* Command and General Staff College, Fort, Leavenworth, KS, B-4. See also Reid, 266. Delbruck, 402, derides the use of modern military terms to describe medieval armies or their actions; however, some organizations and maneuvers are relatively universal regardless of name or nomenclature.

[21] Burne, 238. Harrison, *Joan of Arc*, 149, believes the area was too marshy for siege equipment.

For the English, however, creating the illusion of an offensive siege masked their need for a defensive strategy by serving as an effective cover story for their actual blocking tactic. Military deceptions are designed to capitalize on existing enemy perceptions[22] and if the people of Orleans were anticipating a "siege," why not reinforce that idea?[23] Thus all dispatches referred to the Orleans operation as a siege, but calling it a siege does not make it a siege. This deception neatly concealed an overall strategy designed to hide the fact that English finances were under considerable duress, which strategic deception has also deceived historians.

On the face of it, trying to take Orleans would have been an overly ambitious enterprise requiring massive resources from the English, resources that were not only severely lacking but also considerably dispersed throughout their occupation zone. After all, even Attila the Hun had failed to take Orleans in 451 A.D.,[24] and Henry V notably refused this challenge after surveying the city's formidable defenses in 1421.[25]

The Orleans blocking position also had a domino effect. As stated by the anonymous author of *The Journal of Paris*, "Orleans was still besieged, which made food dearer than ever in Paris, since the town was frequently compelled to send large quantities of flour and other victuals and things necessary for siege warfare to Orleans."[26] Food prices in Paris doubled overnight.[27] Grain prices reached their highest level of the entire fifteenth century in 1428,[28] extending into 1429, making Paris vulnerable to subterfuge while also sapping valuable monetary resources from the English through inflation. Harvesting was usually completed by September 29;[29] consequently, most medieval sieges were designed to prevent this final harvest, thereby denying the sieged population an opportunity to store food prior to the onset of winter. Attacking after the harvest also argues against the idea that the English intended a siege operation at Orleans. It quickly becomes clear that the decision to blockade Orleans was made of necessity not advantage.

22"Deception schemes are greatly assisted by an enemy's self deception." Dewar, 69.

23The outpost of St. Loup may have been an extension of the deception plan to give the appearance of a siege.

24 Eliot, 181; France, I, 120,238; Richards, 74; Purcell, 76; Grayeff, 37; Pernoud and Clin, 12.

25 Tuckey, 17. See also LaCroix and Meller, 87-8.

26 Shirley, trans, 230.

27 Lang, 93.

28 Miskimin, 64,66.

29 Collins and Davis, 26; Nossov, 233.

The tax base also favored France. The estimated population in England at that time was three million compared to a population base of approximately ten to twelve million in France. Meanwhile the gold coinage minted by France between 1428 and 1430 almost tripled that of England. In the case of gold, the years 1424 to 1430 were among the most productive of the century for France, and its silver production massively dwarfed English silver between 1425 and 1430.[30] Of course, with four times the population, the expected circulation of money in France had to exceed that of England by a commensurate amount, but France's silver bullion alone greatly exceeded this ratio. Further, there is no reason to assume that the amount of monetary currency produced in England at that time met the needs of its population; it appears that money was tight.

Thus by escalating this conflict and prolonging England's expensive military operations, Joan of Arc's ultimate relentless aggression openly exposed England's financial weaknesses and placed the duke of Bedford, the overall English commander, overtly, instead of covertly, on the defensive. Although biographers occasionally claim that Joan had no effect on winning the Hundred Year's War,[31] one of her major contributions was to cause a dramatic financial drain on the English by making the war even more expensive.[32] This idea was preached by Dennis Hart Mahan during the nineteenth century (1832 to 1871) at America's military academy, West Point, and practiced during the American Civil War with devastating effect by General Ulysses Grant: "making the assailant [in this case England and Burgundy] share the burdens of war."[33] By making the English pay with men and money, and by making Burgundy pay with the loss of cities and commerce, the duke of Bedford and Philip of Burgundy were forced to suffer ever increasing costs along with their commensurate escalating political stresses, which stresses antagonized Bedford—igniting his obsession to have Joan burned at the stake.

"Bedford himself contributed at least 117,000 *livres tournais* [the name for French currency] of his own money to the siege of Orleans."[34] He also had to enforce salary loans of twenty-five percent on his government of-

[30]See Miskimin, 128,132,136.

[31]For example, see Michelet, *Joan of Arc*, 130, and Goldstone, 187.

[32]Goldstone, 197.

[33]Keegan, *The Mask of Command*, 182. Mahan also emphasized "speed, mobility, surprise" which became the trademark for Joan's military operations. See Tuchman, *Stilwell and the American Experience in China 1911-45*, 16.

[34]Taylor, trans., *Joan of Arc: La Pucelle*, 239n.

ficials and military officers,[35] and threaten six month salary forfeitures if anyone failed to comply with his directive. By fact, the English were so stretched for cash they could not pay their members of Parliament in Paris,[36] which even ran out of money to supply parchment paper for maintaining its official records.[37]

Joan's participation ultimately ramps-up the intensity of the conflict, which extends the war. This intensity continued its relentless increase even after Joan's death; by the end of the war, the English were pawning the crown jewels.[38] Meanwhile for the English, isolating Orleans not only denied its use as a French military base, but also kept the French army otherwise occupied and away from Paris.

The English needed time to consolidate and stabilize their occupation of French lands and then use that occupation to build a war chest for future expansion—to make valid the 1420 Treaty of Troyes. A stalemate at Orleans freed these occupied territories for economic development to increase tax revenues while keeping the French army out of the way. Orleans was nothing more than flypaper in this overall English strategy. The English were fighting for time, the French for a country. Thus the blockade was leisurely enforced, one chronicle describing it as "a great fair, with booths for the sale of all sorts of commodities"[39] all along the English entrenchments. With this summary, we now have a better feel for the actual situation affecting Joan's arrival in Orleans.

According to *The Journal of the Siege of Orleans*, Joan arrived in the city as if God Himself descending from heaven,[40] her life continuing to take on mythological proportions, even as pulpit preachers reported that thunder and lightning presaged her entry into the city. Dunois, the military commander of the French army inside Orleans, had kept his citizens and army informed of Joan's existence and used her progress within the French courts to sustain their morale, not necessarily believing she would make any military difference. Thus Joan's reputation preceded her and the

[35]France, I, 233; Lang, 94; Lucie-Smith, 93.

[36]Michelet, *Joan of Arc*, 45; Michelet, *The Life of Joan of Arc and Other Biographies*, 25; Gower, 82. As further evidence of these dire financial straits, even Cauchon, who conducted Joan's trial, did not receive final payment for his dirty work in 1431 at Rouen until the year 1437. See Pernoud and Clin, 236-37 and Harrison, *Joan of Arc*, 310.

[37]Michelet, 45.

[38]Pernoud, *The Retrial of Joan of Arc*, 19; Pernoud and Clin, 148.

[39]Richards, 77, quoting Andrew Lang, 63, who quotes the Monk of Dunfermline.

[40]Ireland, trans, II, 62; Lang, 118; Beevers, 83; Pernoud and Clin, 41.

citizens mobbed her upon entering the city, but this joy at her arrival was not unprecedented.

During the siege of Toulouse, France, of similar length and seasons and of similar geography to Orleans, sitting on the Garonne River, supplied by boat, and lasting unsuccessfully for nine months from October thru June 1217-18 A.D., the arrival of Raymond VII with copious supplies and reinforcements received almost exactly the same reception as Joan:

> When the people of the town saw [his] banners, they rushed toward the count as if he had risen from the dead....They fell to their knees and kissed his clothes, his feet, his legs, his arms, and his fingers.... Trumpets and horns sounded, the flag-bearers shouted their war cries, and the sound of bells filled the city.[41]

Obviously, those under a siege or blockade experienced serious psychological stresses and naturally met any relief with euphoric celebrations, especially when accompanied by large quantities of food, like the 300 wagons along with the herds of cattle, sheep, and pigs that preceded Joan into Orleans. Experiencing a reception very similar to that of Raymond VII, even setting her banner on fire with their torches, which flames she quickly extinguished with a sharp bolt of impressive horsemanship[42]—to the great delight and wild cheers of her already adoring fans[43]—Joan rode through the joyous crowds of Orleans from east to west, the full length of the city. Nonetheless, despite this impressive reception, the French officers remained in charge and resumed their supply and security operations as normal, without Joan's participation.

Dunois, however, eventually had to leave the city and with the commander absent, Joan again did just what we should expect: she paraded herself through the streets of Orleans like a modern day politician conducting an election campaign and the rapt citizens so flocked to her banner as to crowd the streets and delay her for hours.[44] The Madison Avenue in Joan prevailed. As at Vaucouleurs, Chinon, and Poitiers, Joan went over the heads of resistant institutions and their bureaucracies, in this case the

---

[41]Strayer, 114.

[42]Ireland, trans., II, 62; "Journal of Orleans" in *Joan of Arc: La Pucelle*, Taylor, trans., 356-57; Parr, I, 102-03; Lang, 118; James, 210; Stevens, 92; Richards, 121; Ince, 28; Paine, I, *Joan of Arc*, 142; D'Orliac, 151; Buchan, 103; Beevers, 71; Gies, *Joan of Arc*, 72.

[43]See Pernoud, *Joan of Arc*, 84.

[44]Ireland, trans., II, 179n; Parr, I, 111; Caddy, 184; Adams, 96; France, I, 278; Lang, 122; Bangs, 104; Pernoud, *Joan of Arc*, 85; Pernoud and Clin, 42; De Vries, 78.

military commander and his officers in Orleans, to gain control of popular support, thereby circumventing traditional authority and usurping their powers. Now this is not recommended for every leader, but it has been done before and since in both politics and war.

The Maid represented a brand, a medicinal potion of spiritual revival, which she marketed to a people who very much wanted to believe in it—just like a snake-oil salesman who relies on the placebo effect to validate his brew, thereby continuing its sales. Sometimes *the cure is in the belief.* Joan took the stage inside Orleans and sold her product. When Dunois returned again to Orleans, his city was much different then when he had left it.

With Dunois still gone, Joan also gathered a large number of her Orleans' militia, whom she had whipped into a frenzy of support during her parade through the city, and proceeded to conduct a reconnaissance of the English positions outside the city walls.[45] During this excursion, Joan happened upon some English soldiers posted to keep watch beyond their earthwork. A verbal exchange ensued in which she insisted they surrender. In absolute and utter disbelief an astonished English sentry replied, "Would you have us surrender to a woman?"[46] Joan returned to the city without skirmish and without prisoners.

Nevertheless, once again Joan's *instincts* were true to her profession. She made critical observations and gathered useful intelligence about her enemy by assessing their fortifications, morale, and tactical dispositions.[47] "Know your enemy" is one of the first precepts of war, just as knowing your competition is equally important in business. Further, she had announced and made known her presence on the open battlefield in a fearless manner, surely perpetuating the growing rumors of her supernatural capabilities. Joan must have been a strange sight to see on any battlefield in any age, as with the unique appearances of George Washington, Robert E. Lee, or Chief Crazy Horse on the battlefields of American history. The signature of Joan's plumage had to make for intense scuttlebutt among English soldiers. Joan knew the minds of the common people.

Of interest, Joan's excursion around the English defenses is reminiscent of MacArthur's arrival in Tokyo after the Japanese surrender in WW II.

---

[45] Ireland, trans., II, 65-6; Parr, I, 112; Lang, 122; Waldman, 107.

[46] France, I, 278-79; Bangs, 105; Wilmot-Buxton, 79; Taylor, *Joan of Arc*, 98; Endore, 96; David-Darnac, 63; De Vries, 77.

[47] France, I, xliv, incorrectly believed that Joan "never observed the enemy," which criticism primarily accuses her of not counting numbers and measuring moats.

MacArthur landed in the Japanese capital, stepped off the plane, and strutted about in full view of the Japanese completely unarmed and unguarded. As he expressed at the time, "nothing will impress them like a show of absolute fearlessness. If they don't know they are licked, this will convince them."[48] Joan's brief reconnaissance, strutting about fearlessly in full view of the English, likely had that same effect. Again, she acted with boldness and conviction and never lost an opportunity to take the stage.

Once Dunois returned, military operations began in earnest. A second 300 wagon supply convoy was arriving from Blois, so a second diversionary attack on St. Loup, as had been conducted when Joan first arrived, was needed in order to ferry these supplies into the open eastern city gate. Joan decided to take a nap before this assault but she overslept. Upon awakening, she quickly armored herself and rushed to the city gate to lead her militia into the fray. She was probably an hour late for what was a three hour event.

With an embarrassed bravado, Joan riding frantically toward the eastern city gate to join in the attack against St. Loup must be considered one of the greatest scenes in all of medieval history: sleep in her eyes, white banner snapping her senses awake, mind coming alive to the sights and sounds of a new and forever different life, her rushed armor jostling with loose irritations against the goose bumps of her flesh, even as her anxious spurs poked sharply at the lazy hide of her equally disturbed horse, whose hooves, it was reported, *sparked* against the cobblestone.[49] It certainly rivals her entrance into the grand ballroom at Chinon. This was Joan's moment of metamorphosis, the breakout from her walled cocoon, endless dawns of wrapped evolution bursting forth to acknowledge, in a single moment, one human destiny: "For this task I was born," she had said while in Vaucouleurs. Joan spread wide her fresh sticky wings to fly for the first time in that angelic form of predestined creation. We cannot help but imagine standing on a street corner, peasant hat in hand, cheering wildly as she rides by. Why no artist has ever painted this scene, Joan riding desperately through the city of Orleans with *sparks* flying from the hooves of her horse, is a mystery.

There is a psychological phenomenon called the "grand entrance." Leaders of genius intuitively understand its value; others antagonize us with their pretense. A commander's presence, delayed until the critical

[48]Manchester, *American Caesar*, 444. See also Harvey, 388.
[49]Murray, 296. Testimony of Colette Milet.

time and place of decisive combat, often makes all the difference in soldier morale. Dignitaries sometimes arrive late at social events to signal prestige. A champion prizefighter might "bake" his opponent by arriving late in the ring to make his adversary suffer all the prolonged discomforts of fearful anticipation, those same discomforts soldiers so frequently avoid, as did Joan, by sleeping until the last moments before battle. Wellington, the master of Napoleon, brilliantly timed his infrequent battlefield intercessions to achieve magnificent effect,[50] as did other great commanders.[51]

As fate would have it, Joan's tardiness was a blessing (or evidence of spiritual sanctioning?) because even as the French attacked St. Loup, the English commander unleashed a surprise counterattack against the rear elements of this standard and predictable French diversionary tactic, hoping to trap the French between his two English forces.[52] Without doubt the French were in serious trouble. But exactly amidst this flurry of attack and counterattack, Joan in all her plumage set out from Orleans toward St. Loup with her unexpected and substantial number of French reinforcements.

On that fateful afternoon, the French soldiers attacking St. Loup—about to be ambushed and trapped by the English counterattack, and expecting a frantic fight for their lives—suddenly looked up to catch sight of a single gold-fringed white banner bouncing lightly on the horizon, held aloft by a growing icon of female warrior dressed in white armor, riding a white horse,[53] the only time she was known to use a white horse.[54] She was followed by her loyal brother, a royal aide, a frightened page, and the steaming hot breath of seemingly countless Orleans militia. There can be no doubt that French hearts leapt with joy, and we can be sure they all fell in love

---

[50]At one dangerous moment during the battle of Waterloo, Wellington closed a critical breach in his lines by personally riding forward and leading his troops into the exposed gap. Howarth, *Waterloo: Day of Battle*, 162.

[51]Frederick the Great did the same at a critical moment during the Battle of Zorndorf against the Russians. Koch, 160. Washington rode forward to lead a counterattack against the British at Princeton. Antal, 104. Grant led several regiments forward at the Battle of Shiloh during the American Civil War, inspiring his troops to take advantage of an exposed gap in Confederate lines. Keegan, *The Mask Of Command*, 209. Pershing took over a company in the Philippines to attack a Moro fort "because my presence there was necessary." Tuchman, *Stilwell*, 531. Rommel stepped forward during the Battle for Sidi Muftah in North Africa to take over the lead platoon of the 21st Panzers and break through the British defensive perimeter resulting in the capture of 3,000 allied prisoners. Mellenthin, 125.

[52]Ireland, trans., II, 67; James, 213.

[53]Murray, 30n; Ireland, trans., I, cxliv; Lowell, 98; D'Orliac, 172.

[54]Joan's unusual use of a white horse may validate d'Aulon's testimony that she commandeered her horse from a passing page.

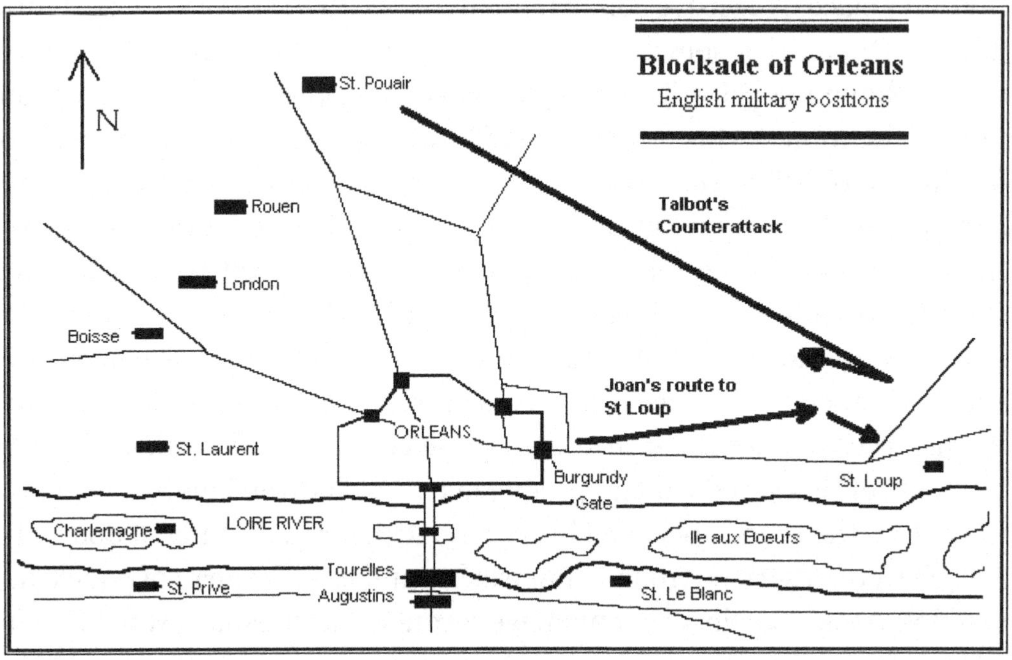

with Joan at that very moment. She was bringing victory to an otherwise sure French defeat.

Joan could not have staged a more spectacular grand entrance onto her first battlefield. The English counterattack, intended to outflank the French concentration on St. Loup, was suddenly forced to turn and fight in two directions or to retreat as Joan's inspired militia flooded the English rear.[55] Caught in a tactical vise, the English forces smartly withdrew,[56] a sight which must have disheartened the 400 Englishmen defending St. Loup.[57]

Returning now to our study of Joan's leadership, John Keegan wrote in his book *The Mask of Command*: "The first and greatest imperative of command is to be present in person. Those who impose risk must be seen to share it, and expect that their orders will be obeyed."[58] When Joan entered the fray at St. Loup, she was joining Dante's eternal flock of butterflies who flutter so defenseless to the Judgment Throne: their grunts and groans, sweat and swearing, cries and cringes, shock and shelling, noise and numbers, dirt and destruction, tears and tyranny, and their fear and their faith.

---

[55]Oliphant, 79; Gies, *Joan of Arc*, 75.

[56]Historical criticisms of Talbot often center on his "impetuous" nature and his "rashness," as well as his "lack of tactical insight" (Pollard, 6), but this St. Loup withdrawal disproves these generalized assessments, assuming he was leading this detachment, which one would suspect, as did Lamartine, 131.

[57]See Appendix A.

[58]Keegan, *The Mask of Command*, 329.

The nineteenth century Clausewitz wrote, "No matter how clearly we see the citizen and the soldier in the same way...soldiers will think of themselves as members of a guild, in whose regulations, laws, and customs the *spirit* of war is given pride of place."[59] The fifteenth century Jean de Bueil, thirty-five years after Joan's death, said it this way: "Those who are not noble by descent are noble by the practice of the profession of arms they follow,[60] which is noble in itself... Arms ennoble a man, whoever he may be."[61] Those who risk life know life differently—just ask the brotherhoods of firefighters and policemen.

"The first and greatest imperative of command is to be present in person. Those who impose risk must be seen to share it, and expect that their orders will be obeyed."

St. Loup was Joan's baptism of fire, her spiritual awakening to a sea of souls unique in history. Although she did little more than participate, it says more of Joan than of most of humanity concerning combat. By all accounts she was nothing more than a cheerleader urging forward her comrades while witnessing their devotion and trust. She never fought on the frontlines at St. Loup and was only a distant target for English archers, but she was a close example to French soldiers. Maybe the French military theorist Ardant Du Picq best identifies why this was so important: "When the battle becomes hot, [soldiers] must see their commander, know him to be near. It does not matter even if he is without initiative, incapable of giving an order. His presence creates a *belief* that direction exists, that orders exist, and that is enough."[62]—*the cure is in the belief.* Joan's presence on the battlefield reinforced that bonding, that primeval spiritual attachment with her soldiers, so necessary to military leadership, or to any leadership, an attachment which began at Blois.

Had Joan arrived in time for this battle her impact on the spirit of the French army, her personal bonding with her soldiers, would not have been as adhesive as it was; her timely arrival changed the battle and her life.

---

[59]Clausewitz, 187.

[60]From General MacArthur: "the soldier who is called upon to offer and to give his life for his country, is the noblest development of mankind." Manchester, *American Caesar*, 172.

[61]Warner, *Joan of Arc*, 160; Keen, 152.

[62]Du Picq, 141.

She was accepted into the soldier's special guild of martial nobility.[63] This psychological phenomenon however was not acquired by chance. Joan's characteristic courage, her willingness to risk her life, is what caused this change, this acceptance. There can be little doubt that courage is a prerequisite to any effective leadership.

Her personal priest testified that after the battle, "Jeanne was much afflicted when she heard that [the English at St. Loup] had died without confession."[64] Joan frequently wept after combat. She saw soldiers, whether French or English, as universal human souls. More significantly, she had witnessed for the first time the true brutality of what only her dreams had imagined: brains smashed and bellies cut open, fingers and limbs severed, men writhing in agony, and...the sounds. Even Wellington, already a veteran commander of sixteen battles and eight sieges,[65] wept at the Battle of Waterloo after defeating Napoleon.[66] But make no mistake, Joan was not against injury or death in war on whatever side, as gruesome as they are, but against death in sin. Catholic confession forgives sin and cleanses the soul to allow entry into heaven. From her perspective, only a pure soul could gather God's intended reward, war or not, and thereby escape the sin of murder. With her priest in tow, Joan went often to the aid of both French and English soldiers for this purpose. Still, witnessing the physical trauma of her chosen profession must have been especially emotional, as it is so often for men. Here then, after the skirmish for St. Loup, we have a perfect example of faith and emotion joining with Joan's intellect and reason.

That night the French officers devised their plans for the next day's action and decided to conduct a full-scale attack on the Tourelles, the two towers guarding the southern bridge into Orleans. This plan called for a diversionary attack on the English outpost of St. Laurent by Joan and her militia, followed by a river crossing and an attack on the outpost of St. Le Blanc by the main French army, with follow-on attacks on the ruined monastery of the Augustins in front of the Tourelles and then on the twin towers themselves.

---

[63] Jean de Bueil also wrote, "A warm feeling of loyalty and pity comes into the heart on seeing one's friend expose his body with such courage to carry out and to accomplish the will of our Creator; and one makes up one's mind to go and die or live with him and, out of love, not to abandon him. No man who has experienced it knows how to speak to the satisfaction that comes from this sort of action." Pollard, 129. See also Wolf, 19.

[64] Murray, 286.

[65] Keegan, *The Mask of Command*, 92.

[66] Ibid., 102; Keegan, *The Face of Battle*, 337.

The next morning, however, the militia rebelled against their role in this plan and decided to go straight for the Tourelles instead.[67] Joan did not lead this attack and remained behind with the main French army while the amateur militia rushed across the river to ransack the English outpost of St. Le Blanc, which the English had quickly abandoned in order to consolidate their forces once they saw the attack coming. Finished with St. Le Blanc, the French militia then moved to attack the Augustinian monastery (Les Augustins). Meanwhile the main body of the French army crossed the Loire River and consolidated behind the militia, watching their action from afar.

In military history, potentially treasonous troops have sometimes been sent first into battle to get rid of them, a tactic practiced as early as Frontinus in the first century A.D., being made known to medieval commanders through the works of Christine de Pizan,[68] and employed as recently as the 1930s by the Chinese General Liu in the Japanese-Chinese war.[69] Thus, as these poorly armed and untrained French citizens, full of spit and vinegar, approached the now heavily fortified monastery, English archers formed themselves forward of the Augustins' bulwarks[70] and let loose their vaunted arrow storm. The impact was instantaneous. The French advance was abruptly halted and the militia thrown into chaotic retreat.[71] This was their first real taste of professional combat and they wanted nothing to do with it.[72] Hundreds of defeated Frenchmen scurried frantically rearward towards the beachhead. The English, inspired by this easy success, cheered their victory and leapt forward to pursue these

---

[67] Murray, 292. Testimony of Simon Charles. Also D'Orliac, 159; Pernoud, *The Retrial of Joan of Arc*, 109; Grayeff, 48; and Pernoud, *Joan of Arc*, 88. The evidence for the rebellion on this day comes from Simon Charles, who again was not present, and who is often less than credible as a second-hand witness to most events. Accordingly, some biographers elect to place this rebellion on the following day, based on Pasquerel's testimony that the French captains had decided not to continue their attack until forced by Joan. However, Pasquerel is an equally unreliable witness. Given the choice between Simon or Pasquerel, Simon seems more plausible.

[68] Pizan, 97-8,101-02.

[69] In battling the Japanese in the 1930's, General Liu, in losing 600,000 troops, calmly stated that it was "really a good thing…The Chinese soldiers are all bandits, robbers, thieves and rascals. So we send them to the front and they get killed off and in that way we are eliminating our bad elements." Tuchman, *Stillwell*, 232.

[70] Ireland, trans., II, 69; Lowell, 104; Lamond, 78.

[71] Lowell, ibid.; Beevers, 80.

[72] The raw Union recruits at Shiloh during our own Civil War also fled in panic upon their first contact with professional Confederate troops. Grant, *Personal Memoirs*, 184.

defeated Frenchmen, fully expecting to inflict heavy casualties as was usually the case in medieval warfare.[73]

By now however many from the French main body of more than two hundred veteran knights and their lances had already crossed the river. Joan rode with La Hire, a French captain who was probably one of the finest soldiers in all of French history, although Michelet calls him a brigand or gangster.[74] La Hire seems to have adopted Joan for this battle. It makes sense; La Hire would have been the one who most appreciated Joan's fortuitous arrival at St. Loup, saving him from behind just days before. La Hire is also the likely reason Joan failed to lead her militia in this attack. He may have convinced her to ride with him because he too knew what would happen to this rebellious rabble. Keeping Joan away from this disaster was another way of thanking her for the St. Loup rescue. Moreover, allowing Joan to ride beside him was tantamount to pinning promotion bars on her shoulders. He signaled to other professional officers his acceptance of Joan as a bonafide leader and legitimate combatant. At the same time Joan could witness valuable lessons in military command: where to locate herself, how to communicate with a strike force, what to watch for, and who to trust. As Joan absorbed these lessons, she could also envision her future role in later French battles.

Joan, sitting beside La Hire, watched as her defeated militia streamed back toward the beachhead, English archers in pursuit. La Hire however calmly dropped his lance into a horizontal position and ordered a counterattack against the English pursuers. This was hardly the flank attack claimed by Michelet;[75] it was a head-on engagement. Joan charged forward beside La Hire while carrying her banner—which banner seems safe enough and certainly not combative. Asked at her heresy trial, "Which did you care for most, your banner or your sword?" Joan replied, "Better, forty times better, my banner than my sword....It was I myself who bore

[73]Lang, 128, Lynch, 150; Monahan, 80-1, and Fabre, 155-54, give another version of events derived either from an unnamed chronicle or from some unnamed source. In this version, which may have been invented by Lang, Joan, together with her militia, supposedly attack the Augustins (which at least solves the problem of Joan not leading her troops) upon which bulwark she finally plants her flag. However, English reinforcements arrive from St. Prive to scare them off. The English then pursue these retreating French, Joan apparently retreating with them, all the way back to La Hire who orders a counterattack against the pursuing English, which Joan joins, reversing her direction, all unlikely. This account by Lang and others somewhat follows d'Aulon's testimony, but with much more detail than his testimony provides.

[74]Michelet, 25.

[75]Ibid., 32.

this banner when I attacked the enemy, to save killing anyone, for I have never killed anyone."[76] Of course military leaders are conductors of death and "Fighting generals have blood on their hands—it is an inescapable part of their profession."[77] Joan's denial of having killed anyone bears witness again to her human frailties—her banner marked the target, inspired the fire, and was every bit as lethal as her sword.[78]

Moreover, throughout military history unit "colors," banners or flags, have been highly prized as war trophies.[79] They almost always draw the most fire, and were often pursued on the battlefield with foolhardy courage by both sides in a conflict. In fact, in their regulations, the famous Templar knights, especially noted for their military prowess, required "from five to ten knights"[80] to pack themselves around their banners to protect them at all costs. Joan had no such protection. During the American Civil War it was said that "Carrying the colors was an almost certain title to six feet of north Georgia soil,"[81] as described during the Battle of Chickamauga. By carrying her banner, Joan marked herself as a special target. Capturing Joan's flag at Orleans would have made any Englishmen a national hero. Although Joan may or may not have been aware of this danger, the soldiers around her would have recognized and respected her courage in facing it, whether she was aware of it or not.

So here we have Joan, whose timely arrival at St. Loup had already inspired her soldiers to victory, now plunging headlong at the front of a counterattack, defying the dangers inherent in her banner, charging fearlessly against a tide of retreating Frenchmen, directly into a wave of victorious English archers. How would any soldier respond to such impressive leadership? The French militia rallied and followed from behind, joining Joan and La Hire's mounted men in the attack.

---

[76]Murray, 31. Nevertheless, there were occasions, such as at the Tourelles and at Paris, when others did carry her flag.

[77]Korda, 356.

[78]Joan claimed to have never killed anyone, but she clearly felt responsible for the English deaths at St. Loup because she found it necessary to make confession after having already made confession that morning. Although Pasquerel claims her confession was because the English dead had not confessed themselves before the battle, there was no way for Joan to know whether each English soldier had or had not confessed.

[79]By example, see Keegan, *The Face of Battle*, 186-88, and Lieven, 134.

[80]Delbruck, 288. See also Lamb, 19.

[81]Tucker, 173. "Much needless loss of life would have been prevented in both armies [in the American Civil War]...if it had not been considered so imperative to carry the colors to the firing line." Ibid.

The English ground forces, now scattered and out of formation, were no match for armored knights on horseback. If the English archers had disciplined themselves and remained massed behind their defensive barriers—most often wooden stakes sharpened and then buried in the earth at an angle that protected the front of their formation, intimidating any attacking armored knights[82]—then their arrow storms might have fended off these French lances, or at least delayed them long enough for a successful escape. Caught two bow shots[83] or 600 yards away from these barriers, however, these English bowmen had no chance. The French knights overran their open retreat and a slaughter ensued with more than 200 killed. The Augustin's defenders were overwhelmed.

Nonetheless these archers were not completely defenseless. They carried with them small weapons called caltrops, multi-spiked iron balls,[84] that when liberally scattered about the battlefield were like a modern hasty minefield serving to puncture the soft center of a horse's hoof, temporarily disabling armored knights.[85] According to Vegetius, the Romans invented these sinister devices to stop chariot attacks.[86] English archers frantically tossed out these weapons as they retreated back toward the Augustins. Unfortunately these nasty little pincushions also hindered human movement and Joan's first combat wound came as a result of stepping on a caltrop in front of the Augustins;[87] tripwires abound and all infantry soldiers learn to watch their step. Joan had dismounted her horse, planted her banner atop the Augustin's earthworks,[88] exposed her "sole," and wounded her foot.

[82]This battle formation was used as early as 1192 at Jaffa where Richard the Lionheart had his knights plant their lances in the ground to protect his crossbowmen as they engaged Saladin's light cavalry. Bridge, 190.

[83]Sowers, trans., *Chronicles of St. Denis*, vii

[84]Sackville-West, 148.

[85]Rankin and Quintal, trans., *The First Biography of Joan of Arc*, 83.

[86]Clarke, John, trans., "The Military Institutions of the Romans (De Re Militari)" by Flavius Vegetius Renatus in *Roots of Strategy*, Philips, ed., 168. However, the Chinese were also using caltrops at about the same time as the Romans. Some were wooden, two and a half feet tall; others were iron and eight inches wide by four inches tall. Sawyer, trans, 77,78. In the East it is said the T'ai-Kung invented them. Ibid, 349.

[87]Rankin and Quintal, trans., *The First Biography of Joan of Arc*, 83; Parr, I, 125; Adams, 104; Oliphant, 83; France, I, 300; Lang, 130; Bangs, 117; Richards, 131; Monahan, 82; Endore, 102; Waldman, 119; Sackville-West, 176; Buchan, 114; Winwar, 105; Leary, 54; Scott, 53; Lucie-Smith, 113; Gies, *Joan of Arc*, 78; Wooster, 104; Maxwell, 80; Richey, 60; Taylor, *The Virgin Warrior*, 69; Harrison, *Joan of Arc*, 154.

[88]Oliphant, 81.

Joan's scene at the Augustins is reminiscent of Stephen Crane's *Red Badge of Courage* where the Union hero, Henry, finally finds his soldier soul and carries his unit colors to victory. This scene is also reminiscent of Douglas MacArthur's father whose heroics at the age of eighteen won him the Medal of Honor when "His faced blackened with smoke, his muddy uniform tattered and blood-stained, he reached the top of the precipice [of Missionary Ridge], and there—silhouetted against the sky, where the whole regiment could see him—he planted the standard [of the 24th Wisconsin at Chattanooga]."[89] And so stood Joan of Arc, her flag planted on the Augustins—daring, visible, and leading the way. During the mounted charge against the Augustins, La Hire validated Joan as a bonafide French officer.[90]

As the sun set on that day's battle, the stage was set for the greatest moment in Joan's history. French boats ferried heavy equipment across the river, troops on both sides repositioned themselves in the dark, the English set fire to one of their outposts, spies sent messages, patrols probed, and commanders weighed their options as decisions frayed the nerves on both sides. The common soldiers ate and drank their fill, and slept the heavy sleep of stuffed stomachs, avoiding the anticipation of combat.[91] Joan rode into Orleans to have her foot treated and to get a good night's sleep. She would not awake the next day to construct a battle, but to join it. She would not influence combat preparations, but advantage their results. She would not just follow, but command. And she would not just watch and cheer, but bleed.

The final attack on the Tourelles would be the moment of truth in Joan's history, the knockout round, the proof of her righteousness or the utter humiliation of all she believed. As Joan slept that night she lived her battle a thousand times and was ready. As she proudly testified, "It was I who first planted a ladder against the fortress of the [Tourelles]."[92]

The English were about to experience an even bigger surprise.

---

[89] Manchester, *American Caesar*, 14.

[90] Junior officers were often designated to carry the unit colors, their baptism by fire so to speak. Keegan, *The Face of Battle*, 186.

[91] "We will make ourselves comfortable and sleep, and eat as much as we can stuff into our bellies, and drink and smoke so that the hours are not wasted. Life is short," (Remarque, 87), which compares favorably with "Let us eat and drink; for tomorrow we shall die." Isaiah 22:13.

[92] Murray, 32.

Chapter Five

HER LEADERSHIP IN VICTORY

Previously we set the stage for Joan's entry into Orleans by reconstructing English strategy and redefining their military operations as an effort to create a defensive buffer zone around Paris. At the southern edge of this buffer zone, Orleans was blockaded to tie up the French army and keep it away from the capital. We then witnessed Joan's joyous reception into the city and her subsequent parade through that city to acquire, once again, public support. We then watched as she exposed herself outside the Orleans walls in front of the English positions, fearlessly announcing her arrival to the English soldiers. Although late for her first combat, Joan led a courageous counterattack against an English effort to relieve St. Loup, thereby establishing her leadership credentials with the French commanders and their soldiers. We also acknowledged her emotions and faith after this battle. Finally, we rode with her as she attacked into the teeth of an English victory, carrying her banner, reversing the battle outcome, and taking the Augustins monastery. Thus her leadership at Orleans exampled brazen *audacity* inside and outside the city walls, *courage* and daring in two life threatening conflicts, and *humanity* in the face of brutality. But her real leadership talents are about to burst forth in all their glory—after all, she has been virtually immortal for 600 years, and here is why.

May 7, 1429, was not a normal day in the history of France. Joan awakened to her legacy, armed herself for combat, gave confession, attended mass, and set out for immortality. As the sun rose, she crossed the Loire River to join the French army at the Augustins.

As the other French commanders gathered at the frontlines, Joan participated in all those sensations known only too well to all soldiers rousing themselves for early morning combat. Stiff bones from shivering sleep cracked with the heavy breath of fogged air; exertions increased and heartbeats raised the pulse of life; the whites of eyes appeared and perspiration beaded foreheads; quivering arms tightened straps, sharpened blades, and adjusted loads; earthly movements changed from sluggish dreams to alert alien jerks; and tired minds gave way to darting pupils. Men in artillery batteries checked their cold guns and sighted their weapons. Leaders paced these busy lines, supervising soldier preparations with watchful eyes. Slowly a mass soul formed to test the wares of war in readiness for life or death. The warface of man painted across an army.

Soldiers coming awake feed on themselves, especially so for early morning combat, as if an orchestra warming its instruments, plucking their quiet rustlings into a crescendo of raucous commands, hearty words, tough demeanor, and back-slapping bravado, until the cacophony ends with the conductor tapping his podium and raising his baton—the staff of his command. Then silence engulfs the stage even as the first cannon balls whistle overhead. A battle-ready army plays as crisp and clean as a bugle's first call.

The arrival of Joan's banner, flapping briskly in the early morning breeze, likely emboldened these French souls, undefeated, having caught the English twice in the open and routing them both times. Superstitions always exist and Joan's appearance likely signaled good luck. Maybe she really was from God. Likely soldiers cheered her arrival; likely their inspiration spurred her foolishness. Victory is always bittersweet and Joan's glory was also Joan's defeat.

At this point many biographers place Joan in command.[1] The idea is preposterous. Despite overwhelming evidence of heavy logistical preparations as proof of a fully intended assault, many biographers accept a single witness and his testimony that the Orleans' civil magistrates arrived early that morning to convince Joan of the need to countermand fictitious orders from the French captains not to attack. Likely it was the militia who stirred up these rumors, if such an event ever took place. In defiant response Joan supposedly mounted her stallion and shouted "Let who loves me follow me!"[2] This proclamation however is about as far out of character for Joan as anyone can imagine. Given what we know of her motives and character,

[1]For example, see Fuller, *Decisive Battles*, 487.

[2]Parr, I, 127; Endore, 105. "Follow me!" is the motto for the American infantry. See also John 21:19: "Thus spake he, signifying by what death he should glorify God. And when he had spoken this, he sayeth unto him, 'Follow me.'"

she could have cared less for, nor would she have ever invoked the idea of, personal loyalties. She was fighting for a higher mission on God's behalf and if anything wanted soldiers to follow Him.[3] We can dismiss outright the idea that Joan took command the morning of May seventh, or that she forced the French commanders to attack.[4]

Further testimony suggests that Joan rode directly to the Augustins, immediately commanded the bugles to sound, and with banner waving led the attack.[5] All such stories make for great theater and are scenes we would all like to believe and embellish, but if Joan was ever real, the truth tastes more of mud.

La Hire had validated Joan as a French officer, but she failed to understand that cherished responsibility, not to fault Joan whom we should hardly expect to comprehend such military honor. Instead she was of another vision, one we should expect: the vision of a frontline soldier. As often happens with new officers, they seek approval through sacrifice and frequently represent themselves as brave, foolhardy, and daring. A good commander identifies such careless souls and keeps them off point until reality sets in, lest they become needless casualties.[6] In this regard La Hire failed at his command, because he failed to recognize in Joan that recklessness of character.[7]

> **"U.S. Army Doctrine: Leaders identify subordinates who exhibit high-risk behavior. High-risk behavior is a behavioral pattern that intentionally or unintentionally increases the individual's probability of negative consequences."**

We can also assume that the French assault on the Tourelles actually began at Dunois' command. Most likely the previous night's military council had designed a coordinated attack plan that sent all units simultaneously

[3]"She ascribed all to God, and, so far as she could, resisted when the people sought to honour her or give her the glory." Murray, 249. Testimony from seven burghers of Orleans.

[4]Nevertheless, this version of events is often accepted. For example, see De Vries, 82,84.

[5]For example, see Lynch, 153.

[6]Even high level commanders act recklessly. Eisenhower had to admonish Patton, "I want you as Corps Commander—not as a casualty," to keep him away from the frontlines. Bradley and Blair, 184-85.

[7]Army Leadership, 6-6.

against the bulwarks and walls of the Tourelles, with bugles signaling the advance,[8] thereby forcing the English to dilute their defenses and protect everywhere at once. The Tourelles was a short distance from the Augustins so the time of exposure to reach the protective bulwarks and the tower walls was minimal. Thus, somehow, in the middle of all these attack preparations, Joan managed to disengage herself from La Hire, team with a ladder, and at the bugle's call charge the two towered Tourelles with a band of frontline infantrymen. Because of the short distance she must have been on foot, a sore foot, and could not be stopped. Nevertheless at this instant Joan of Arc became the first ever bonafide "infantrywoman."[9] Thus, as soldiers scurried like turtles,[10] and as English defenders discharged their parapet weapons, Joan scrambled forward to make history...and was instantly wounded.

Joan knew not whether she was hit by an archer's arrow or a crossbowman's bolt, but the projectile struck hard between her neck and breast.[11] Thus, within the first few moments of battle, Joan was out of action and inconsequential to further combat. Nevertheless some biographers place her atop a ladder when wounded,[12] and some have her aloft waving her banner,[13] but such images are legend. Was Joan really going to engage seasoned English professionals in hand-to-hand combat atop a high precipice? Would she really waste a good ladder by blocking access so she could wave her flag? Extremely doubtful. Joan testified, "It was while raising this ladder that I was wounded in the neck."[14]

---

[8]Murray, 316. Testimony of d'Aulon. (Delbruck, 310n, in a book that covers medieval warfare from 634 to 1522 A.D, claims he could not "remember reading in any medieval source anything about signals in battle." He then unexplainably lists several instances of such usage, but skips entirely the French campaigns under Joan of Arc. D'Aulon in fact references the use of trumpets during the battle for the Tourelles at Orleans.)

[9]See Appendix B.

[10]Soldiers in siege assaults used shields, but not the type typically carried by knights on horseback. These shields were shaped like half barrels, each with two straps that could be strung under the arms to keep the hands free, and were worn on the back to protect against arrows, stones, and burning oil. Assaulting soldiers looked very much like turtles with long ladders protruding from beneath their shells—heads and buttocks protected. France, I, 366; Sackville-West, 148; Leary, 56.

[11]Murray, 32.

[12]Lamartine, 132; Creasy, 219; Stanhope, 47; Bartlett, 140; Adams, 106; James, 69; Whetnall, 106; Haggard, 286; Stevens, 103; Taylor, *Joan of Arc*, 125; Stolpe, 123; Trask, trans., 176.

[13]Taylor, ibid.; David-Darnac, 71; Maxwell, 89.

[14]Murray, 32.

Joan was wounded because she wore plated armor, sectioned in order to allow for freedom of arm and shoulder movement.[15] Consequently, when she raised her arms to assist with the ladder, these sections separated and exposed gaps. Although Joan likely remained protected by a thick undergarment, at a distance of less than thirty feet, with gravity assisting, a strongly pulled bow or normal crossbow could easily penetrate this safeguard.

Again trial testimony disputes the exact events surrounding Joan's wound. Dunois, the Orleans commander, testified, "She continued nonetheless to fight, taking no remedy,"[16] but he may have been confusing his memory of Orleans with that of Paris where Joan did remain on the battlefield after being wounded. On the other hand, her priest testified that the arrow (or bolt) "went through from one side to the other."[17] Even Dunois claimed it penetrated six inches.[18] Such a wound had to be temporarily incapacitating, so she must have been taken from the battlefield as described by her priest and others.

Joan had no business leading this attack—except to fulfill a previous prophecy in which she predicted her wound. Joan, as *Chef de Guerre*, was fully out of position and inappropriately deployed and thereby lost to her army. As a young lieutenant, however, Joan may have been exactly where she belonged, leading her soldiers.

Notwithstanding, there is a terribly strange aspect to warfare understood by Stephen Crane and by Joan, which remains valid even while contradicting military professionalism. Theoretically a perfect battle results in no wounds and no deaths,[19] but a perfect soldier carries *The Red Badge of Courage*, the Purple Heart, a battlefield wound. It is a paradox, irony, and frustration of war that wounds sanction a soldier's bravery. As stated by Keegan in *The Face of Battle*, it is "the receipt of wounds, not the infliction of death" that demonstrates an officer's courage.[20] Yet, by all reasonable measures battlefield injuries are tributes to fundamental mis-

---

[15]See Lang, 107, and Gies, *Joan of Arc*, 59.

[16]Murray, 236; Sowers, trans, *Chronicles of St. Denis*, vii.

[17]Murray, 290; Ireland, trans., II, *Journal of the Siege of Orleans*, 70. Taylor, *The Virgin Warrior*, 70-72, by retranslating, claims the arrow actually "passed by her," and that Joan suffered only a glancing flesh wound. Joan then conducted a ruse to make it appear she had been seriously wounded. Of course to believe this one has to ignore Pasquerel and Dunois' testimony and accept the idea that a ruse takes almost three weeks to heal. Murray, 32,71.

[18]Murray, 236.

[19]"To subdue the enemy without fighting is the acme of skill." *Sun Tzu*, Griffin, trans., 77.

[20]Keegan, *The Face of Battle*, 191.

takes. Nevertheless, the soldier *ethos* attributes these catastrophes to an uncontrolled fate; after all, even the very best soldier doing everything right can still be wounded or killed. Therefore any wound however acquired becomes a badge of honor and a legendary war story.

By all accounts, La Hire, the very best commander and soldier in the French army, was never wounded at Orleans, but Joan, a teenage female, needed that revered credential and foolishly over-exposed herself to achieve it.

Joan's armor was removed for medical treatment and, according to her priest, a soldier offered enchantments to ease her pain, but Joan refused believing such conjurations evil.[21] The priest further testified that Joan wept and made confession.[22] In the midst of all this trauma, Joan says the vision of her saints arrived to comfort her.[23] She spent the rest of the day recuperating.

With Joan absent, the French commanders assaulted the Tourelles numerous times without success.[24] This fierce resistance by the English belies the idea, often seen in chronicles and in modern biographies, that Joan's reputation as a witch always inspired an overwhelming fear in English soldiers.[25] Even when heavily outnumbered, as was most often the case during the French Loire campaign, the English fought with admirable tenacity.

As evening approached, Joan heard the French bugles sound retreat. The battle was over. According to Dunois, the hour had reached eight p.m.[26] and darkness was upon them.[27] It must be remembered that when Joan entered Orleans at eight p.m., little more than a week before, the townspeople had met her with torchlights.[28] Dunois called it a day and recalled his army. And so would end Joan's story: the little girl who had tried but failed.

---

[21]Murray, 288.

[22]Ibid.

[23]Ibid., 32.

[24]Bangs, 123.

[25]For example, whenever Joan waves her banner the English flee in fear. Bray, 133.

[26]Murray, 236; Leary, 57.

[27]D'Aulon testified, "they were before the Boulevard from morning till *sunset* without being able to take it or gain it." Murray, 316. Simon Baucroix, a squire who was present, testified, "the attack lasted the whole day, up to *nightfall*. . . .When *evening* came. . .another attack was made on the fort." Ibid, 268. De Contes testified, "The king's troops remained there [the Tourelles] from morning to *night*." Ibid., 262; Lucie-Smith, 119. *The Chronicles of France* says the final attack occurred "about sunset." See also Sowers, trans., *Chronicles of St. Denis*, vii; and the "Journal of Orleans" in *Joan of Arc: La Pucelle*, Taylor, trans., 358, which says from "morning until sunset."

[28]Murray, 236. Testimony of Dunois.

But Joan's legacy was at stake. She envisioned French victory. This was the predicted round of triumph. This time Joan raised her wounded flesh and frantically rushed to Dunois begging him to delay the army's withdrawal. "Rest a little, eat and drink," she protested.[29] She then forced herself into the saddle and rode off to a nearby vineyard to pray: her garden of Gethsemane—"The spirit...is willing, but the flesh is weak."[30] Fifteen minutes later,[31] Dunois says half an hour,[32] Joan returned from the vineyard inspired—and in command!

The real proof of Joan's miraculous spirit begins with her impalement on the cross of the Tourelles. Her blood and sacrifice struck a religious chord in these Christian soldiers. When she returned to the battlefield resurrected, her wound treated with olive oil and lard,[33] this symbolism was not lost on the mass souls of her endangered comrades.

Like it or not Joan was imaging the life of Christ: unique to the world, flamboyant in character, unremitting in purpose, declarative of supreme authority, sacrificial in blood, resurrected in kind, and predestined for immortality. Equally important, as Keegan points out in *The Face of Battle*, Joan's "insistence on returning" to the battlefield after being wounded "consummated" her honor, and it is "by establishing one's honor with one's fellows that leadership is exerted indirectly over the common soldiers."[34]

By example, when General Leonard Wood remained on the battlefield with a shattered leg at Lovejoy's Station in the Georgia campaign of the American Civil War, General Sherman remarked "that his example of courage was worth 20,000 men."[35] By returning to the battlefield, Joan instantly won the unequivocal respect and the universal loyalty of her army.

Joan of Arc followed a different destiny, separate of human reason—a destiny of faith. Before the battlements of the Tourelles, Joan, with her wound, proved to herself that she was an instrument of God, a great conductor of life's orchestra. And every French soldier understood that message as they witnessed her miraculous resurrection from the grave. Presumed

[29]"Journal of Orleans" in *Joan of Arc: La Pucelle*, Taylor, trans., 357; Parr, I, 133; Oliphant; 86; Sackville-West, 181.

[30]Matthew 26:41.

[31]France, I, 309; Lang, 137; Waldman, 126; Fumet, 28; Winwar, 109; Leary, 57; Pernoud and Clin, 47; Taylor, *The Virgin Warrior*, 71.

[32]Murray, 236; James, 70.

[33]Murray, 288-89. Testimony of Pasquerel.

[34]Keegan, *The Face of Battle*, 191.

[35]Tucker, 252.

dead, she was no longer just a good luck symbol, or just another commander; she had become an ephemeral spirit, an unfathomable force of victory. That moment, when Joan reentered the field of combat, was the instant of her immortality. As so eloquently stated by Karen Sullivan in her study of Joan of Arc, "Unlike the ability to swing a sword or plan an attack...the ability to restore morale that Joan exemplified exists only at the moment that its possessor is recognized as legitimate."[36] Joan's wounded sacrifice confirmed her legitimacy and cloaked her in the mantle of command.

With Joan's resurrection, she became a leader of men, men who shared her soul, men of mud, men of blood, and men of a battle-tested spirit who have forever honored *The Red Badge of Courage.* Winston Churchill spoke of "blood, toil, tears, and sweat" to save England during World War II. It is no accident that he spoke first of blood. He understood Joan of Arc, the savior of France, and just days after Christmas in 1938 wrote to his wife, Clementine, saying of Joan: "I think she is the winner in the whole of French history. The leading women of these days were more remarkable and *forceful* than the men."[37] He recognized Joan's strength of will.

It is no small achievement to raise combat veterans—fatigued, hungry, aching, and disheartened—from a bugled retirement to risk their lives yet again for the seeming impossible, asking them to overcome unimaginable miseries and imagined defeats, and once more challenge the mysteries of fate. But Joan had invested her soul in this army and she called on that spiritual capital to bet it all on this one moment. She could ride forward to utter defeat, soldiers laughing in derision, or raise the hearts of men to immortality.

To the French commanders, Joan was not in charge, but to French soldiers Joan commanded their faith. They had seen her riding with La Hire. And she had never lost. And she was willing to attack. And she had nurtured their will to fight. And she had been mortally wounded. And she was resurrected. And so...they rose with her and to her calling, like the exhausted Maine Regiment on Little Round Top at Gettysburg who saved the American union with a final bayonet charge because they had run out of ammunition or the impetuous British Light Brigade who followed Shakespeare's immortal words as put in the mouth of Henry V to charge "once more unto the breach."[38] Unequivocally, the final French attack on

[36]Sullivan, 79.

[37]Manchester, *The Last Lion*, 432.

[38]Harrison, *Henry V*, Act III, Scene I, line 1, p.747.

the Tourelles came only at Joan's command, and it was her instant of eternal history.

Joan returned from the vineyard, rode forward alone, planted her banner high atop the enemy bulwarks, and commanded her soldiers to strike once more, to attack blindly into that black abyss, and in so doing inspired her army to victory.[39] Now this was some theater, almost unequalled in history.

Napoleon stated, "In war there is but one favorable moment; the great art is to seize it."[40] Clausewitz said, "War consists...of single, great decisive actions, each of which needs to be handled individually."[41] Patton believed, "Before an attack is declared hopeless, the senior officer must lead an attack in person."[42] In person and decisively, Joan seized the favorable moment—the "great art."[43] As recognized by Sun Tzu, "When the strike of the hawk breaks the body of its prey, it is because of timing."[44]

The startled English defenders trembled in disbelief as they witnessed the entire French army suddenly rise up and surge forward—yet again—with another massed assault against their besieged fortress, this time led by a ghost. Joan's final summons for yet another French attack released all the spirited ferocity of resurrected souls rising to the bugled promise of a fresh new dawn. The stunned English instantly abandoned their positions, unable to match the inspired French legions now led by Joan's unwavering will to win.

And here is the critical point: it appears that most of Joan's men rose to fight in that final assault. The numbers of soldiers do not mean anything unless those numbers actually participate. As du Picq recognized, "The Romans did not pay so much attention to [numbers]. What they paid most attention to was seeing that everybody fought."[45] He went on to say, "The theory of strong battalions is a shameful theory. It does not reckon on courage but on the amount of human flesh. *It is a reflection on the soul.*"[46] Of course the soldier's soul was always Joan's primary concern when lead-

---

[39]Murray, 236,247. Testimony of Dunois and Jean Luillier.

[40]Philips, ed., "Military Maxims of Napoleon" in *Roots of Strategy,* 436.

[41]Clausewitz, 153.

[42]Axelrod, 108.

[43]Delbruck, 326, reasoned, "The united *will* that has to govern the whole body was not provided by the organization of the [medieval] army as such, which decides everything according to the *will* of the commander, but it had to be created and guaranteed for each occasion."

[44]*Sun Tzu,* Griffin, trans., 92.

[45]Du Picq, 131.

[46]Ibid.

ing her troops into battle, which gave most if not all of them the courage required to follow her commands and to rise up and fight when called upon in this final assault.

But where could the English flee after their bulwarks were breached? Glasdale, the English commander, had been fighting from an earthwork construction in front of the Tourelles[47] and his only hope for escape was into the Tourelles by way of a small drawbridge and then out the back toward the river bridge, which bridge had been previously cut by the French. However the French were now attacking from that direction also. Wooden planks attached to drain pipes[48] had been thrown across the broken arches by which the French militia, from inside Orleans, were also advancing against the two towers. The English were being surrounded.

Thus, still facing the maniacal French army now swarming the bulwarks, Glasdale withdrew toward the smaller drawbridge and its entrance into the two towers. It seems however that the French had anticipated this route of retreat also, evidence that Dunois had his own thoughts of victory from the beginning. A barge filled with "pitch, tow, faggots, horse-bones, old shoes, resin, sulfur, and ninety-eight pounds of olive oil"[49] had been drifted and beached beneath this wooden drawbridge leading to the two towers—and then set afire.[50] English troops faced a gauntlet of soldiers, weapons, and fueled flames.

This fire ship however has often been overlooked by historians. It adds to the timeline necessary for Joan's success. Although the French had been ordered to retire, they must have been making ready this barge. It also takes time to drift and lodge this ship and then to have it burn through the wooden bridge. Either the bridge was already burning when the French attacked, or there was enough time to execute this diabolical ploy. Apparently

[47]"The English [at the Tourelles]...arranged all things, such as canon, crossbows, stick slings, culverins, stones, and other implements of war *necessary for their defense*, in case they were assailed." Ireland, trans., II, *Journal of the Siege of Orleans*, 9. It is important to realize that this structure was defensive in nature, further supporting the idea that the English were not offensively minded, at least tactically, which also hints at their overall strategy. See also De Vries, 61.

[48]Ireland, trans., II, *Journal of the Siege of Orleans*, 72; "Journal of Orleans" in *Joan of Arc: La Pucelle*, Taylor, trans., 357; Parr, I, 135; Leary, 58; Lucie-Smith, 120.

[49]France, I, 311; Winwar, 110; Fabre, 159. See also Parr, I, 130.

[50]Adams, 110; Whetnall, 108; Bangs, 125; Richards, 139; Taylor, *Joan of Arc*, 131; Paine, I, *Joan of Arc*, 174; Endore, 107; Leary, 58; David-Darnac, 72; Sermoise, 81; Grayeff, 57; Pernoud and Clin, 47; De Vries, 89; Richey, 61; Harrison, *Joan of Arc*, 157.

an enterprising fisherman had managed to beach this barge behind enemy lines and make a miraculous escape.[51]

One must also question the northern attack across the broken bridge span. Was this a coordinated participation, or an *ad hoc* maneuver as described by Michelet?[52] Was it designed by signal, or coincidental? It seems too much to believe that all of this happened within minutes and by accident. There had to be some method to this madness, which would require time and planning, especially to pre-construct the bridge pieces. This final attack had to be more deliberate than most historians suspect, and all of these calculations had to derive from a leader. Joan and Dunois did more than simply order another assault. Most likely a prepared contingency plan was implemented that caught the English napping. There is some evidence to support this theory if we remember Joan's promise that morning to return to Orleans *by way of the cut bridge*.[53] If Joan actually made this prediction, she either knew of the expected bridge attack as part of the original plan, or she had it in mind to implement this maneuver. If so, it suggests that the militia were in fact left inside the city to protect it, while also being assigned a supporting mission, possibly on signal, to attack the Tourelles from along the bridge—a secondary or diversionary attack of the type originally planned against St. Laurent. If this design came from Joan, then one can certainly understand why the French commanders began to praise her generalship. After all, she was a quick learner and knew of the planned diversion in the original attack plan, and "vicarious learning can be as effective as first-hand experience."[54] Further, one witness testified that Joan had previously ordered the "small gate near the great tower"[55] opened that morning, which suggests she may have been entrusted to control that element of an already planned operation.

The persistent idea, based on a few assertions made by early historians and mindlessly copied over time, that medieval commanders did not

[51]From the ledgers of Orleans is found this entry: "To Jean Poitevin, a fisherman, eight *sous* for having beached a *chaland* which was put under the bridge of the Tourelles to fire them when they should be taken." Sackville-West, 185; Harrison, *Joan of Arc*, 159. See also Paine, I, *Joan of Arc*, 345n.

[52]Michelet, *Joan of Arc*, 34-5.

[53]See also Lang, 351-2n. Pernoud and Clin, 47, claim that the soldiers themselves were promising to return by the bridge and they credit Pasquerel, Joan's priest, with this testimony. By fact the statement "In God's name, tonight we shall enter the city by the bridge," actually came only from Joan based on testimony from d'Aulon, not Pasquerel. Murray, 317.

[54]Buonomano, 136.

[55]Murray, 262.

understand the art of war,[56] nor employ any form of tactics other than a disorganized melee, simply does not hold up when professionally scrutinized.[57] Clearly they understood the tactical use of deception, feints, diversions, demonstrations, and psychological warfare, as well as the use of scouting, security, screening, special operations, echeloned and staggered attacks, maneuver, and reserves. They also understood the importance of patrols, ambushes, raids, and surprise to gather intelligence, effect counterintelligence, interdict supplies, or to disrupt enemy operations. They organized effective medical evacuation and combat communication systems, engineered static and mobile obstacles, and used both artificial and natural terrain to their advantage. They often established and exercised a chain-of-command, even if subordinate commanders sometimes felt justified in attempting independent initiatives, not always successful. And sure, they held staff meetings from which they derived their plans, but what modern army does not? They also understood force ratios and the need to shape the battlefield to influence those ratios, and the need for logistics planning to support the priority of effort. And they designed broad strategies, integrated with military operations to support political goals. Joan's ultimate contribution was to inject speed and violence—both a surprise to the English—onto this traditional battlefield. In fact, George Bernard Shaw felt that she was the "first French practitioner of Napoleonic realism in warfare."[58] Patton believed that the golden rule of warfare was speed, simplicity, and boldness.[59] Although this brief period of history may be an exception to the rule, it seems doubtful. Viewed from the foxhole, these soldiers and their commanders were more universal than most historians like to give credit.

Legend has it that Joan revenged her prior English insults by yelling to the English commander, Glasdale, "Surrender to the King of Heaven. Thou hast called me a whore but I have great pity on your soul and the souls of your men!"[60]—again Joan was concerned with the spirit and not the body. But Glasdale and his knights refused to surrender and instead chose the

---

[56]For example, see Gies, *Joan of Arc*, 85

[57]The idea that commanders in the Middle Ages "were mere pawns of circumstance" has been discredited. "Medieval military historians long ago disproved these misconceptions." Hay, 78-9, gives a good summary of the sources available to refute this persistent fallacy.

[58]Shaw, "Saint Joan" in *Joan of Arc: Fact, Legend, and Literature*, Jewkes and Landfield, eds., 168; Raknem, p.193.

[59]Axelrod, 135

[60]Endore, 107; Sackville-West, 184.

gauntlet of fire sweeping the drawbridge. Unfortunately, by the time of their attempted escape, the bridge had been so charred and weakened by the flames burning beneath that it collapsed under their heavy burden.[61] Glasdale and his men fell through the flames and into the water where they drowned in the weight of their armor.[62]

The remaining approximately six hundred English, caught inside the Tourelles and trapped by the collapsing drawbridge, were methodically eliminated or captured.[63] Overall about one hundred Frenchmen were killed or wounded.[64] If this number is true, it is testimony to the timidity of the day's fighting without Joan.

That night Joan returned to Orleans across that same southern bridge that had been blocked by the English for almost seven months.[65] She had captured it in little more than seven days. Her enemies were not entirely defeated but Orleans was saved and the way opened to Reims. This young teenage girl from Domremy, bleeding from her chest, was no longer human. She had risen from the dead, commanded an army, and defeated the greatest adversary in French history to that time. She would never be viewed the same again. Even La Hire must have shaken his head in disbelief.

---

[61]Ireland, I, clviii; in the "Journal of Orleans" in *Joan of Arc: La Pucelle*, Taylor, trans., 358. Michelet, *Joan of Arc*, 35; Creasy, 221; Stanhope, 49, Bartlett, 142; Parr, I, 134; Bray, 138; Haggard, 287; Winwar, 110, and Trask, trans., 178, all accept that the drawbridge was collapsed by cannon fire coming from the French artillery positioned on the river island, which seems equally plausible as a contributing factor and which testimony comes to us from the squire, Simon Baucroix (Murray, p.268), but without any other corroborating testimony. It is possible that both artillery fire and the barge fire weakened and then destroyed the bridge.

[62]Supposedly, an astrologer had advised Glasdale that he would never be harmed by cannon, fire or iron; no mention was made of water however. Endore, 107.

[63]Sowers, trans., *Chronicles of St. Denis*, vii; Charles VII letter to Narbonne in *Joan of Arc: La Pucelle*, Taylor, trans., 86; France, I, 312; Paine, I, *Joan of Arc*, 348n. Bray, 138, Adams, 111; Lynch, 155, Creasy, 221, and Trask, trans., 178, report three hundred killed and two hundred captured. The troops from St. Prive had been added to the original 500.

[64]France, I, 312-13n, gives an excellent summary of sources and their various casualty figures for this battle, ranging from four hundred to 1,000 English slain or taken, against only five to one hundred French casualties. He ignores Monstrelet, 553, who states that 6-8,000 English were killed or taken, clearly a misprint, misinformation, or an exaggeration. See also Joan's testimony, Murray, 32.

[65]However, the bridge would likely have required repairs first; the initial makeshift crossing was described as a gutter or drainpipe that was "marvelously long and narrow, and high in the air, without having any support." Ireland, trans., II, *Journal of the Siege of Orleans*, 72; Waldman, 126; Lucie-Smith, 120. Accordingly, the *Chronicles of France* (Sowers, trans., *Chronicles of St. Denis*, vii) says that she slept in the field because the bridge across the Loire River was still broken. Nonetheless, subsequent events bear out that she more likely returned to Orleans.

Wordsworth comes to mind:

> A perfect woman, nobly plann'd
> To warn, to comfort, and command;
> And yet a Spirit still, and bright
> With something of angelic light.

Across the Land of Gaul blew a breath of fresh air. Spring flowers bloomed brighter and more fragrant along the yellow-brick road to Reims. Butterflies spread their new wings, coming alive to a world revived and more beautiful. Church bells rang with victory instead of warning. And a new future dawned as a new name swept the land—"The Maid of Orleans!"[66]

The English belief in Joan's supernatural powers really began with her victory at Orleans, not because of her unlikely command of French troops, but more likely because of the conditions conspiring to enhance that command. We know enough about her wounds, faith, and courage as fundamental to her battlefield success, but a blind faith night assault as accomplished under Joan's inspired leadership is the most frightening military tactical circumstance any soldier will ever face. As Caesar noted in his *War Commentaries*, "It is, indeed, generally the case that people are much more frightened of what they cannot see than of what they can."[67] Modern analysts agree:

> Night has been commonly characterized as "no man's friend." Surrounded by darkness, people tend to imagine sinister forces lurking in quite harmless objects; every unknown sound seems ominous. Perceptions become distorted: objects appear larger than life, and distances appear greater and are more difficult to calculate. The psychological toll this can exact, when coupled with hunger, fatigue, and combat excitement, can engender near-panic or even mass hysteria among frontline troops.[68]
>
> In most of the races of man, superstition unfolds with the night, tricking the imagination and stifling courage.[69]

---

[66]This sobriquet was first published in 1552, and some claim did not exist during Joan's lifetime; however, it seems so obvious a name that it must have survived as verbal history, much like the *Song of Roland*, long before finally placed in writing—an act so necessary to scholars—the written word finally catching up with oral history. See Pernoud, *Joan of Arc*, 105.

[67]Warner, trans. *War Commentaries of Caesar*, 179.

[68]Sasso, Claude R. *Leavenworth Papers No. 6: Soviet Night Operations in World War II.* (Fort Leavenworth: Combat Studies Institute, 1982), 5.

[69]Marshall, *Pork Chop Hill*, 176.

At Orleans, Joan transformed the medieval battlefield with a mania for victory that speaks of satanic possession. To fight in the dark is to face the devil.[70] There is nothing more heart stopping than to confront a belligerent professional soldier in ebony darkness who wants to bayonet your heart and take your soul. Night combat is the ultimate test of the will to fight. Current technologies reduce the psychology of this trauma for the American soldier, but for those combatants still lacking night-sight technologies, the modern battlefield remains primitive, evil, and devilish.

Joan led blind men against blind men at the Tourelles. The fact that her troops were at all aggressive, fiercely attacking, made all the difference.[71] Henceforth, for the English there could be no sanctuary, no rest, no time for comfort, no recovery or temporary escape. The setting sun no longer offered a reprieve from soldier suffering because the witch of the night, now terribly feared, owned the psychology of the battlefield.

Combat at ground zero is very much a bumbling chaos, what Patton called "an orgy of disorder."[72] Too much attention to detail renders a commander useless. Leaders must be visionaries who keep in mind their broad objectives, exercise their necessary support structures, and command their most destructive systems while still managing to understand the most frightening effects on their own soldiers who are forever trapped in this maze of autocracy—it has always been so.[73] Joan swinging her sword is wasteful; Joan climbing a ladder is fruitless; but Joan rising to the intense passions of her desired military objective, invoking her physical example to inspire the martial spirit, and bleeding her exhausted flesh to achieve success, despite the clump, thump, and frump of dirt, dark, and disaster is the epitome of command.[74]

To French soldiers the consequences of combat became secondary to the overwhelming demonstrated vision of their female leader. Joan saw victory where others knew defeat because they could not see, and she changed that

---

[70]Even in L. Frank Baum's *Wizard of Oz*, 111, the wicked Witch of the West is afraid of the dark.

[71]According to Wolf, 228n, the Battle of the Herrings allegedly began three hours after midnight, which, if so, would demonstrate an earlier French propensity for accepting the feasibility of night attacks.

[72]Atkinson, *An Army at Dawn*, 136.

[73]"Leaders must understand stress and anticipate Soldiers' reactions." Army Leadership, 9-3.

[74]"He who does not feel strong enough to keep his heart from ever being gripped by terror, should never think of becoming an officer." Du Picq, 120.

blindness[75] with her supremely sighted and sanctioned *will.*[76] It is this aspect of Joan's victory at Orleans that historians rarely recognize but which makes her a bonafide military commander.

What has been said of the sixteen-year-old military genius, Charles XII of Sweden, can also be said of Joan: "The *firmness* with which he opposed misfortune, his *indefatigable activity* in all his enterprises, and a *heroic courage* which was *blind to danger* were certainly the characteristic traits of this extraordinary monarch."[77] Chartier, the king's secretary, described Joan's leadership characteristics as: "military prudence...*courage*...*action*... justice...virtue and joyfulness. She is graced with these qualities beyond any other."[78] Clearly Joan was in her element when leading the French army.

George Bernard Shaw, awarded the Nobel Prize for literature in 1925, defined genius as a "person who, *seeing farther* and probing deeper than other people, has a different set of ethical evaluations from theirs, and has *energy* enough to give effect to this extra vision and its valuations in whatever manner best suits his or her specific talents."[79] It was Joan's greater vision that made her a genius and it was the energy of her will, her special talent, that brought that genius to life. Will Durant, winner of the Pulitzer Prize in 1968 and awarded the Presidential Medal of Freedom in 1977, defined genius as, "an individual endowment of force and will, but it is also a social heredity of *discipline* and *skills* formed in time and absorbed in growth."[80] Besides her forceful will, Joan's inherited discipline and her obviously rapid growth in military leadership have been previously demonstrated.

Gabriel Hanotaux in 1911 identified four unexplained mysteries in Joan's life. The first mystery was "the formation in her mind of the call to such unwavering perseverance."[81] Clausewitz, in his discussion of military genius felt that *strength of will*, determination, perseverance if you will, could not

---

[75] "She saw so clearly—the world moved so slowly—and she might die before she made it see." Royden, 14.

[76] "In military as in other human affairs will is what makes things happen." Tuchman, *Stilwell*, 561. We must also remind ourselves that this was not the first documented victory attributed to Joan's strength of will. In part, the Poitier's council approved her for battle because of "her unceasing perseverance and the urgency of her plea." Stevens, 80; Beevers, 54. See also Antal, 33.

[77] Luvaas, ed. and trans., *Frederick the Great and the Art of War*, 344.

[78] Paine, I, *Joan of Arc*, 362n.

[79] Shaw, "Saint Joan" in *Joan of Arc: Fact, Legend, and Literature*, Jewkes and Landfield, eds., 169.

[80] Durant, *The Reformation*, 223.

[81] Stevens, 54.

exist without some measure of doubt from the commander, which doubt Joan seemed to exhibit by riding into that vineyard. Clausewitz further explains, "Determination, which expels doubt, is a quality that can be aroused only by the *intellect*,[82] and by a specific cast of mind....It is engendered only by a *mental act*; the mind tells man that boldness is required, and thus gives direction to his *will*....This particular *cast of mind*...is what makes strong men determined."[83] He went on to emphasize, "Men of low intelligence...cannot possess determination in the sense in which we use the word. They may act without hesitation in a crisis, but if they do, they act without reflection; and a man who acts without reflection cannot, of course, be torn by doubt."[84] Joan's brief hesitation—presumably praying in the vineyard—before ordering the final French attack meets all the requirements set forth by Clausewitz for a *determined* military genius and begins to unravel some of Joan's mystery.

In his further considerations of military genius, Clausewitz went on to discuss the need for strength of character[85] because of the various unexpected effects that enemy actions have on a commander. He directly assessed the impact of an "ebbing of moral and physical strength" in an army and of a commander's subsequent responsibilities. He concluded,

> The inertia of the whole gradually comes to rest on the commander's *will* alone. The *ardor of his spirit* must rekindle the flame of purpose in all others; his inward fire must revive their hope. Only to the extent that he can do this will he retain his hold on his men and keep control.[86]

Field Marshal Irwin Rommel, known as the Desert Fox for his ingenious exploits in North Africa with the German army during World War II, believed:

---

[82]Stevens, ibid., also makes this interesting observation: "Intelligence is insufficient to be wise to a distant purpose, and it is never reached as originally willed, or found to be worth the value that first inspired the way, but *intelligence is always sufficient for the process of perseverance*, in which life is a development of the infinite moral system."

[83]Clausewitz, 103.

[84]Ibid.

[85]As stated by General Archibald Wavell, "[A general] must have 'character', which simply means that he knows what he wants and has the courage and determination to get it." Harvey, xliii.

[86]Clausewitz, 104-05. "All Army members will experience situations when it would seem easier to accept defeat rather than finish the task. During those times, everyone needs an inner source of energy to press on to mission completion. When things go badly, a leader must draw on inner reserves to persevere." Army Leadership, 4-2.

> No army is composed solely of heroes. They will insist for one reason or another that they cannot go on. The commander with his authority must combat these natural phenomena of weariness, and wrench his officers and men out of their apathy. The man [woman?] in command must be the galvaniser of the battle. He must be constantly on the battlefield, in the front line, to exercise his control.[87]

The ardor of Joan's spirit rekindled that flame in her army, and her inward fire galvanized their courage. She took control, thereby also fulfilling this criteria for military genius.

Clausewitz concluded, "Such are the burdens in battle that the commander's *courage and strength of will* must overcome if he [she?] hopes to achieve outstanding success."[88] Between the Athenian's defeat of Persia at the Battle of Marathon in 490 B.C. and Wellington's defeat of Napoleon at Waterloo in 1815, Joan's victory at Orleans has been considered one of the fifteen most decisive battles in military history[89]—in accordance with Clausewitz, "an outstanding success."

Concerning courage, Clausewitz went on to say,

> Since danger is the command element in which everything moves in war, courage, the sense of one's own strength, is the *spiritual* factor that influences judgment. It is the lens, so to speak, through which impressions pass to the brain....But courage is by no means a conscious act; like fear, it is an *emotion*. Fear is concerned with physical and courage with moral survival.[90]

Joan brought that moral *emotion* to the physical fears of her army.[91] She was no longer human, neither to her adversaries nor to her compatriots. She had stepped out of that mundane world of corporal anxiety and entered the mystical realm of emotional valor, that spiritual orb of moral unknowns, the numinous universe of immortality, and the amalgamation of clairvoyant dreams—the true composition of real mud and the hardened mortar of her castle walls.

---

[87]Carell, 185.

[88]Ibid., 105. In assessing Rommel's character and leadership, Mellenthin, 57, stated, "What I admired most were his *courage* and resourcefulness, and his incredible *determination* under the most adverse circumstances."

[89]Whetnall, 83; Richards, 140; Rousseil, 60.

[90]Clausewitz, 137-38.

[91] "Understanding that *emotional* energy sparks motivation and endurance is a powerful leadership tool." Army Leadership, 5-2.

Further, as du Picq reminds us, "...soldiers themselves have emotions. The sense of duty, *discipline*, pride, *the example of their officers and above all their coolness*,[92] sustain them and prevent their fear from becoming terror."[93] In his book *Battle Studies*, Ernest Judet defined courage as "the domination of the *will* over *instinct* even in the greatest danger."[94] Against all human instinct, Joan charged forward first, posted her banner atop the bulwarks in front of the Tourelles, and stood with a wounded coolness, bleeding beneath the bugle's moonlit call, setting that example for French soldiers.[95] Mark Twain recognized this military gift when he said "...there is that in her heart that raises her above men...and turns mobs of cowards into armies of fighters that forget what fear is when they are in that presence."[96]

But the real core of Joan's success may have best been reasoned by George Bernard Shaw in *Back to Methuselah* when he recognized that "You imagine what you desire: you *will* what you imagine; and at last you create what you *will*."[97] Modern explorations of neuroscience (the chemical functioning of the brain) in leadership confirm Shaw's observation: (1) "Focus is power. 'The act of paying attention [the *will*] creates chemical and physical changes in the brain'," and (2) "Attention density shapes identity. 'Repeated, purposeful, and focused attention can lead to long-lasting personal evolution.'"[98] Joan's focus on her objectives, her *will* to create the France she envisioned, was never absent.

"You imagine what you desire: you will what you imagine; and at last you create what you will."

Nevertheless, had Joan lived after Martin Luther and had she been able to read the Bible, her less secular explanation might have been, "Faith is

[92]"Effective leaders are steady, levelheaded when under pressure and fatigued, and calm in the face of danger." Army Leadership, 5-3.

[93]Du Picq, 120.

[94]Ibid., 18.

[95]"The first of all requisites for a man's [women's?] success as a leader is, that he be brave. When a general is animated by a truly martial spirit and can communicate it to his soldiers, he may commit faults, but he will gain victories and secure deserved laurels." Jomini, 345.

[96]Searle, 37.

[97]Ibid., 111.

[98]Matthew E. May. "The Neuroscience of Leadership" on the website Innovation Excellence (http://www. innovation excellence.com/blog/2011/12/09/the-neuroscience-of-leadership/), December 9, 2011.

the substance of things hoped for....in His word do hope....because the sword of the spirit is the word of God."[99] Of course Joan's voices were that word and her resulting spirit was that sword. There are no better descriptions for how Joan of Arc functioned.

Why did Joan attack again after Dunois ended the battle? What happened when she retreated into that vineyard? We do not know, because no one asked her. Did her subsequent actions come from her heavenly voices or from her human *will*? Did Joan sense something in the English body language and voice inflections that triggered her intuitions? Andrew Lang made the general observation that "Unlike the generals with whom she rode, she divined the temper of the enemy; she foresaw how they would behave,"[100] which is not unlike Fuller's description of Caesar who also "divined his enemy's intentions and...set aside his own fears."[101] Did Joan intuitively understand that the English were vulnerable and thereby *reason* her decision to renew the attack even before praying for divine support? Joan told her troops to rest a little, but the word "rest" implies a temporary pause, as with an interval of silence between musical notes, which does not imply an end to the music. Had Joan already decided on another assault even before entering that vineyard? If so, her mysterious voices had nothing to do with her decision. As darkness approached, did Joan detect or suspect something in the English reactions to the French retirement that told her they were ready to quit, that she could surprise them?[102] If so, we have at least some clues as to her military genius, and to her *intellect*.

"Faith is the substance of things hoped for....in His word do hope....because the sword of the spirit is the word of God."

But let us enter briefly into the English foxhole to remind ourselves of those other soldiers, and to realize more practically the true circumstances experienced by these warrior souls. When Dunois signaled his retirement,

[99]Hebrews 11:1; Psalm 130:5; Ephesians 6:17. Stevens, 54, acutely observes: "faith is the way of life, and will is the way of individual judgment."

[100]Lang, 5.

[101]Fuller, *Decisive Battles*, 199.

[102]Maurice Saxe, an eighteenth century Marshal of France, made some interesting observations about deciphering various "signs" unwittingly made by the enemy, which could be used to take advantage of various weaknesses. See Philips, trans., "My Reveries on the Art of War" by Marshal Maurice de Saxe in *Roots of Strategy*, Philips, ed., 292-94. See also *Sun Tzu*, Griffin, trans., 118-22.

either by bugle, banner or voice command, the English commanders likely did the same. By all reasonable expectations the fight was over because night had fallen.[103] Thus the English soldiers collapsed to rest, took off their helmets, exchanged brief tales, moved to eat or sleep, and otherwise began the process of ending their day.[104] There seemed plenty of time to stand down.

In the interim Joan roused herself, located Dunois, rode off to pray, and then returned to the field. As much as half an hour had easily passed. By this time the English mind and body, so willing to fight all day long, had lost its focus and vitality and relaxed to enjoy a few hours of safety. English soldiers accepted their success, slacked off (in the modern soldier's vernacular), and turned to other pursuits.[105] The few sentries left on the ramparts and bulwarks for security would have also become less diligent. Then suddenly distant bugles echoed, war cries sounded, and masses of French troops swarmed the ramparts, "as though a flight of birds were settling on a hedge."[106] Naturally the English were caught off guard.

It is also important to recognize that these English soldiers inside the Tourelles had been relatively sedentary for seven months and the training necessary to sustain their mental toughness and combat readiness appears to have been lacking.[107] At most artillery gunners had been kept in fighting trim. Moreover it is one thing to risk one's life when prepared to do so; it is another thing altogether when unprepared. Xenophon, writing at about the same time as Sun Tzu (circa 400 B.C.) half a world away, noted: "The unexpectedness of an event accentuates it, be it pleasant or terrible. This is nowhere seen better than in war, where surprise *terrorizes* even the strongest."[108] In essence, Joan's renewed attack ambushed the English,

[103]"a defensive battle that remains undecided at sunset can usually be considered a battle won." Clausewitz, 530.

[104]"An army is never less disposed to fight than immediately after a victory, when everyone shouts for joy, each exaggerates his great feats of arms, the multitude is delighted to have passed successfully out of great dangers, and nobody has any desire to face these dangers in the field." Luvaas, ed. and trans., *Frederick the Great and the Art of War*, 331.

[105]"During the early morning spirits are keen, during the day they flag, and in the evening thoughts turn toward home. And therefore those skilled in war avoid the enemy when his spirit is keen and attack him when it is sluggish and his soldiers homesick. This is control of the moral factor." *Sun Tzu*, Griffin, trans., 108.

[106]Lowell, 109; France, I, 310; Fabre, 159.

[107]"An army in which vigilance is not perpetual is sick until the enemy demonstrates it to be dead." Du Picq, 9. Patton confirms, "Trepidation and loss of morale come with a defensive posture." Axelrod, 60.

[108]Du Picq, 261.

caught them off guard, *terrorized* them mentally, and thereby changed the course of history.[109]

Further, as du Picq recognized, "A man surprised, needs an instant to collect his thoughts and defend himself; during this instant he is killed if he does not run away. The surprised adversary does not defend himself, he tries to flee."[110] As one witness testified concerning the Tourelles, "nor did [the English] during this attack even defend themselves."[111] Clearly the English had been surprised and, according to J.F.C. Fuller, "one of the main pillars of generalship....[is] to do something that the enemy does not expect, is not prepared for, something which will surprise him and disarm him morally."[112]

Whatever the case, there were no military councils when Joan returned from her assumed prayers in the vineyard. There were no debates. Whatever Joan said to Dunois, if she said anything at all, was decisive and went unchallenged. She took *command*.

English soldiers did not fear Joan so much as fear the spell of her moral influence. They had been surprised and shocked by the mad faces of confrontational Frenchmen hung high above an invisible abyss inspired to night craziness, unrelenting, irrational, and savage. How could such Frenchmen, so easily exhausted by sunlight, become so demonic in moonlight? For the English the answer was clear, the witch—Joan of Arc.[113]

Some might argue against this portrayal of Joan's victory as nothing more than an accident of time and space, like the unpredictable frictions of a rollicking roulette wheel whose little white marble has landed several times in a row on the same number. Yet we might also argue as did Napoleon that "War is composed of nothing but accidents."[114] In fact, Clausewitz observed that "No other human activity is so continuously or universally bound up with chance. And through the element of chance,

[109]Far off in Italy, a clerk of Cremona also recognized that surprise was critical to Joan's Orleans' success when he wrote: "The victory may be explained by an advantageous position and an unexpected attack." France, I, 384.

[110]Du Picq, 43. "Fear can paralyze a Soldier. Fear of the unknown can be terrifying." Army Leadership, p.9-4.

[111]Murray, 263. "The commander should, if he sees the opportunity, attack his enemy...when they think they are safe, because those surprised in such a way neither skill nor strength will help them, nor are numbers of any use." Pizan, 46,46n.

[112]Fuller, *Generalship*, 32.

[113]Night was believed to be the time for witch activities. Russell, *Witchcraft in the Middle Ages*, 23.

[114]Philips, trans., "Military Maxims of Napoleon," in *Roots of Strategy*, Philips, ed., 436.

guesswork and luck come to play a great part in war."[115] Eisenhower wrote to Marshall before the invasion of North Africa stating, "To a certain extent a man must merely believe in his luck."[116] We might conclude then that most victories in war are in fact accidents or just plain luck in the sense that what actually happens is rarely predictable. Patton's code name for Third Army during WW II was "Lucky,"[117] nevertheless he kept a vision for each intended engagement upon which he constructed his plans and ordered his operations knowing full well that he would have to react with flexibility to the realities of the battlefield.[118] Joan of Arc was much the same, especially toward her brief maturity as a commander.

Some biographers contend that Joan's military popularity was based on soldier superstitions and that she was viewed as nothing more than a good luck charm, as if a rabbit's foot in our modern pocket.[119] In fact, Alice Buchan stated bluntly that she believed Joan's military ability was "three parts luck and one part straight common sense"[120] (Michelet's influence again). To the contrary, most soldiers simply prefer winners because they save lives and end their travails. Joan may have been lucky at times, but she was usually prepared for that luck, which is often what military command is all about. Evan Connell, in his book about Custer, a commander who also believed openly in luck, got it right when he said, "Luck may be just luck, as everybody knows, yet more often it is the result of ability, *instinct*, or knowledge."[121] Then again, "A bold general may be lucky, but no general can be lucky unless he is bold."[122]

The next day, during the early morning mists, Talbot, Scales, and Suffolk, who had not centered their commands inside the Tourelles and therefore remained alive, and who had withdrawn their English forces from their other earthworks, now consolidated their remaining 4,000 English for

---

[115] Clausewitz, 85.

[116] Atkinson, *An Army at Dawn*, 67.

[117] "In the allied battle code each of the armies had a secret name. Most of them were high-sounding. SHAEF was called 'Liberty,' Twelfth Army Group was 'Eagle,' First Army was 'Master.' Patton prophetically named Third Army 'Lucky.'" Hatch, 173.

[118] Napoleon: "What is luck? The ability to exploit accidents." Esposito, unnumbered page introducing the "Italian Campaigns." According to Pizan, 79: "Chance gives victory more often than does strength." According to Machiavelli "three things are necessary in war: many good soldiers, prudent generals, and good fortune." Bondanella and Musa, eds. and trans. *The Portable Machiavelli*, 312.

[119] France, I, 266, 372-73; Endore, 472; Guillemin, 74; Scott, 226n.

[120] Buchan, 258.

[121] "Custer believed in luck, which may not be irrelevant." Connell, 132.

[122] As said by General Archibald Wavell. Harvey, xliv.

battle in the open field, hoping to entice the French into a foolhardy attack against their massed archers.[123] Dunois responded boldly, positioning his army to face the English. The other captains inside Orleans rode out to join this French formation. Joan arrived wearing a light coat of chainmail,[124] such was the sensitivity of her deep wound. For over an hour the two armies faced each other in patient confrontation until, unlike before previous battles, the French commanders consulted Joan, "Should we attack?"

With a master's touch Joan did not answer. Instead she ordered her army to hear mass. A marble table with Catholic accoutrements arrived on the battlefield where the French proceeded to celebrate the mass of St. Michael in full view of the English.[125] These French victors of a crazed night assault turned their backs on the English and knelt to worship. Once again Joan prepared her soldiers for combat by strengthening those human channels leading to *emotional* courage. Nevertheless, at one point during these ceremonies, Joan asked, "Are their faces or backs toward us?" The English were retreating. "Let them depart," she commanded.[126]

There can be little doubt that Joan staged this battlefield mass. It is quintessential Joan, especially when denied her costuming for visual effect whether such flare was premeditated or instinctual. But the most convincing evidence that she deliberately staged this theater derives from its enduring impact on English soldiers who later rumored that during the final battle for the Tourelles they had witnessed St. Michael on a winged horse riding across the night sky aiding the French attack.[127] English minds joined their two most embarrassing moments: the defeat at the Tourelles; and their order to retreat from a kneeling army during the Feast of St. Michael. Joan was also a genius of symbolic psychology.

However, with this decision, Joan again established another set of military credentials. Roger Nye, in his book *The Challenge of Command*, arrived at four criteria for becoming a *warrior*: first, "to be courageous in the face of great adversity;" second, "to ensure that...soldiers fight with courage and aggressiveness;" and third, "to wage war with the violence necessary to achieve victory." Certainly Joan satisfied these criteria, but Nye surprisingly adds a fourth and final requirement which Joan now

123 Murray, 247. Testimony of Jean Luillier.
124 Ibid., 297. Testimony of Aigan Viole.
125 Ibid., 249. Testimony of Jean de Champeaux.
126 Ibid.
127 Michelet, *Joan of Arc*, 35; Stanhope, 48; Bartlett, 141; Bray, 137; Adams, 110; James, 77; Sackville-West, 183; Winwar, 110; Fabre, 161.

fulfills. Warriors must also "restrain their use of violence" when appropriate.[128] More than any courageous conduct of violence, we may reason that an appropriate restraint of violence is what adds that final critical ingredient so necessary to honor, so necessary to the profession of arms. It is this nobility of restraint[129]—"Sweet mercy is nobility's true badge," said Shakespeare[130]—that finally merits *warrior* status. As recognized by the modern American army, "Ambiguous situations, such as when to use lethal or nonlethal force, are a test for the leader's judgment and discipline. The Warrior Ethos helps create a collective commitment to succeed with honor."[131] Accordingly, with Joan's commitment to save soldier's lives, on both sides, and to allow an English retreat,[132] it now seems appropriate in all the debates about Joan of Arc to classify her as both a military *commander* and a soldier *warrior*, in accordance with both modern and historical professional standards.

In fact, although we might be accused of piling on, Robert Harvey added a fifth criteria after comparing a dozen famous generals, a criteria that Joan also fulfills: "Clearly, a degree of madness was helpful to the achievement of ultimate warrior status."[133]

There are some few authors who scorn Joan's reputation as a military leader. Therefore, we must now claim to have refuted George Eliot, who felt she did not possess "military genius," and Francis Leary, who felt she was not "truly a warrior," and Vita Sackville-West, who felt she was not a "professional soldier."[134] Clearly they are wrong because they were blind to the criteria necessary to argue their point.

Most important to our considerations, however, is the extent to which Joan stacks up against the "Great Captains" of history. Actually, that question is not that hard to answer. Using the generalship qualities Liddell Hart believed most distinguished these men, we find an immediate comparison

[128]Nye, 91.

[129]From Thucydides: "Of all manifestations of power, restraint impresses men most." See also Powell, 207.

[130]Harrison, *The Tragedy of Titus Andronicus*, Act I, Scene I, line 119, p.298.

[131]Army Leadership, 3-4 and 5.

[132]During the campaign for Lerida (50 B.C.), Caesar had made this same decision, writing "Why should I lose any of my own men even in Victory? Why let them be wounded when they had served me so well? And why tempt Fortune? After all, a good general should win victories by intelligence just as much as by brute force." Nevertheless, his decision not to fight was unpopular with his men. Warner, trans. *War Commentaries of Caesar*, 246.

[133]Harvey, xx.

[134]Eliot, 189; Leary, 59; Sackville-West, 149.

with Joan: "a blend of *acute observation* with swift-sure *intuition*; the ability to create *surprise* and throw the opponent off balance; the *speed* of thought and action that allows the opponent no chance of recovery; the combination of *strategic and tactical* sense; the power to win the *devotion* of troops and get the utmost out of them."[135] Observation, intuition, surprise, speed, action, military prudence, and devoted troops, all define Joan of Arc. After studying twelve of history's most famous commanders, Robert Harvey arrived at sixteen traits common to all great leaders. In abbreviated form the first five were; (1) outstanding and exemplary courage; (2) the ability to think coolly and rationally; (3) remarkable determination; (4) extraordinary stamina; and, (5) resolution in the face of setbacks.[136] Joan fulfills not only these five, but almost all of the other remaining eleven traits as well.

Three virtues at which Richard the Lionheart excelled while in his twenties were: "supereminent valor and daring; unbounded liberality and bountifulness; and steadfast constancy in holding to his purpose and his word."[137] Joan of Arc again. Napoleon declared, "What I am I owe to strength of *will*, to character, application, and audacity."[138] Joan could say the same. Patton believed that the qualities of a great general were tactical aggressiveness, strength of character, steadiness of purpose, acceptance of responsibility, energy, and good health.[139] Enough said on behalf of Joan.[140]

In assessing the leadership qualities of his combat commanders during WW II, Eisenhower made a very unusual but insightful observation. He said,

> Commanders are habitually diffident where they are called to deal with subjects that touch the human soul—aspirations, ideals, inner beliefs, affection, hatreds. No matter how earnestly commanders may attempt to influence a soldier's habits, his training, his conduct or extol the virtues of gallantry and fortitude, they shyly stop short of going into matters which they fear may be interpreted as preaching.[141]

---

[135]In the forward to Guderian, 15.

[136]Harvey, xlvi.

[137]Bridge, 57.

[138]Durant, *The Age of Napoleon*, 243.

[139]Axelrod, 28.

[140]Hay, 244, described the qualities of Matilda of Tuscany as "illustrating the effectiveness of urban alliances, maneuvre, surprise, interior lines, combined arms and Fabian tactics [war by attrition]," which sounds like a combination of Joan and Charles VII.

[141]Eisenhower, *Crusade in Europe*, 454.

As unusual as this may sound, Eisenhower was tapping into a very delicate but universal military deficiency, as recognized by David Thomas in his introduction to *Hellenika*, written by Xenophon more than 2,000 years before Eisenhower. He said, "The place of psychology in changing the course of battles is a central preoccupation for Xenophon and something he emphasizes far more than do other ancient historians, perhaps because his personal military experience was especially great and varied."[142]

Like Eisenhower and Xenophon, Joan of Arc understood, whether with premeditation or instinctively, the psychological needs of the battlefield, and she stands out in that regard, even among some of our own nation's greatest commanders. She was not shy about the military art of inspiring the soldier's soul. She was not afraid to be accused of preaching.

If the military profession requires what is called "military art," then we have to recognize what art is—bringing the intangibles of life to life. When we ignore these intangibles, "aspirations, ideals, *inner beliefs*, affections, hatreds," we ignore the greater part of the military profession, the part that in fact makes it an art...and creates success. As General William Tecumseh Sherman recognized in his personal memoirs of the American Civil War, "There is a soul to an army as well as to the individual man, and no general can accomplish the full work of his army unless he commands the soul of his men, as well as their bodies and legs."[143]

No matter how advanced technology becomes, even if war evolves into nothing more than machine against machine controlled by toggle switches, there is always a human hand behind it, responsible to build it, employ it, fight it, and account for its outcome. Joan took the machinery of war as it existed and applied art to its use. She was not afraid to preach and she was not shy about reaching out to touch the human soul.

If there is a single leadership lesson in the purpose of Joan's life, it is in the importance of engaging the intangibles of life within the human soul. This is the leadership that overcomes obstacles and allows each of us to face—even death.

On the day of the English retreat, the Orlean's City Register recorded, "the raising of the siege of Orleans was the greatest miracle in history since the time of our Lord."[144] More miracles were yet to come. With the battle

---

[142]Xenophon, *Hellenika*, p.xlii. Xenophon consistently regarded tactical ability "as less important than creating good morale." Ibid.

[143]Sherman, 387.

[144]Rousseil, 33; Lamond, 84; Rutter, 174; Lightbody, 85.

over but the war unfinished, France had found itself a new commander, a new warrior, a military genius, destined not only to annihilate the English army in the field, fighting eleven more battles, but also destined to crown Charles as the true king, a king who would ultimately define the parameters for a new nation....France.

EPILOGUE

To secure the crown of France, Charles VII needed to secure his route to Reims,[1] and to secure his route to Reims he needed to make safe the cities inside the Bourges salient. His army would have to abandon that region during its ride north, so pushing back the English defenses was a necessary preparation,[2] for "it is a rule of war that [an army] must never make deep penetrations...without protecting its rear and providing for its security."[3]

On June 9, 1429, a month and a day after her victory at Orleans, Joan returned to that city where the townspeople greeted her as might be expected, flocking to her as if a modern rock star. Nonetheless, Joan had returned not for adulation but to fight again. This time her objective was the walled city of Jargeau with its five towers and three gates, just twelve miles east of Orleans. Jargeau was also surrounded by a moat and like Orleans had a bridge across the Loire River. Approximately seven to nine hundred English defended the city,[4] slightly more than the number who had been inside the Tourelles. Thus, on June 12, another French army of maybe 8,000[5] strong marched to array itself before the walls of the city.

---

[1]Reims was the traditional site for French coronations because it possessed the sacred oil necessary to sanctify French rulers.

[2]Guillemin, 101.

[3]Lavas, ed. and trans., *Frederick the Great on the Art of War*, 350.

[4]Ireland, trans., II, *Journal of the Siege of Orleans*, 82; Sowers, trans., *Chronicles of St. Denis*, ix; Parr, I, 154; Caddy, 206; France, I, 349; Lang, 156; Bangs, 147; Lynch, 161; Paine, I, *Joan of Arc*, 210; Endore, 120; Waldman, 143; Rutter, 182; David-Darnac, 78; De Vries, 103; Richey, 67; Taylor, *The Virgin Warrior*, 78; Goldstone, 139; Harrison, *Joan of Arc*, 171. Leary, 62, says, 1,200; Lowell, 115,124, says 5-600 English; and Richards, 164, says 600.

[5]Ireland, trans., II, *Journal of the Siege of Orleans*, 82; Gower, 64; Lowell, 124n; Julleville, 56; France, I, 347; Lang, 152; Endore, 120; D'Orliac, 173; Rutter, 188; Stolpe, 140; Burne, 250; Leary, 63; David-Darnac, 78; Richey, 67; Reid, 277. Richards, 160, ignores the militia and says only 3,000. Gies, *Joan of Arc*, 92, feels 8,000 is an exaggeration and suggests 3,600 to 4,800. Oliphant, 98, and Fabre, 178, estimate 3,600 regulars plus militia. Pernoud and Clin, 59, and Taylor, *The Virgin Warrior*, 78, estimate 4,000. Lowell, 123,124n, uses 4,000 of which only 3,000 are regulars. Lamond, 93, and Beevers, 93, estimate 5,000. Lang, 157, eventually reduces his numbers to 5,600, while Waldman, 143, starts with 2-3,000, grows it to 5,000

As at the Tourelles, the French spent the night and all the next day making probing attacks without decisive results. French cannon softened up the walls and three towers were dropped during the shelling. Much of the military praise for Joan comes from her perceived abilities to deploy and understand artillery. One French commander testified, "In assembling an army, ordering military operations, [and] directing artillery, she was most skillful. It was above all in making use of artillery that she was so wonderful."[6] This time Joan ordered the final attack and with the English heavily outnumbered the battle was brief. After four hours, she rode forward to the city wall, climbed a ladder with her banner in hand,[7] and was hit on the head with a spent stone that broke apart on her helmet.[8] She fell to the ground only to rise and yell, "They are ours! Keep a good heart!"[9] The city was taken.

Ignored by historians and biographers but of considerable importance surrounding Joan's attack on Jargeau is the nature of her aggressive assault on this city. The French officers had previously debated their military strategic and tactical options against the English and Joan had disagreed with them and with the king's final decision. However, once that decision was made and the orders issued, she wholeheartedly executed the king's

---

(ibid.) and ends with 7-8,000 (147). Monstrelet, 553, and Paine, *The Girl in White Armor,* 111, say 5-6,000. Goldstone, 139, estimates 5-8,000, while Lynch, 161, and Harrison, *Joan of Arc*, 173, estimate 6-7,000. Finally, Caddy, 206, and Whetnall, 116, place the number at 7,000. Paine, I, *Joan of Arc*, 207, solves the problem by stating 3-8,000. Adams, 121, and Pernoud, *Joan of Arc*, 113, use Alencon's testimony that there were 1,200 lances, which forces various calculations. If the size of a lance is three, then the number is 3,600; but if the lance is five, the total is 6,000; Adams then adds 2,400 bowmen to this number. Nevertheless, 8,000 seems reasonable because the army eventually shrinks to 5,000 after Jargeau and balloons again to 12,000 shortly before the campaign to Reims, meaning that Joan's successes result in a combined 7,000 volunteer militia, of which 2-3,000 early arrivals seem reasonable for the Jargeau campaign, while the regular French force maintained its usual 4-5,000. Further, a large part of the militia volunteers were likely used for logistics support and not as fighters, depending on the availability of weaponry.

[6]Murray, 281; Rankin and Quintal, trans., *The First Biography of Joan of Arc*, 139, suggests that Joan learned her artillery skills from "Ambrose de Lore and others like him."

[7]Lang, 158; Bangs, 149; Paine, I, *Joan of Arc*, 210. This event, which may have been invented by Lang, may be why some biographers place Joan on a ladder with her banner when assaulting the Tourelles.

[8]Murray, 278; Ireland, trans., II, *Journal of the Siege of Orleans*, 84, and Studer, trans., *Mystery of the Siege of Orleans,* xxix, say an Englishman cast down a hard stone. "A piece of armor that may have belonged to Joan is the bassinet (shallow helmet) now displayed in the Metropolitan Museum in New York." Parr, I, 158; Pernoud and Clin, 224; Castor, 115.

[9]Murray, ibid.

plan. This kind of loyalty (as long as the plan is legal in accordance with the times) is an officer's duty and critical to the discipline and order of an army. Disagreement never means disloyalty. Joan's adherence to this essential aspect of military professionalism—a willing obedience to legal orders—lends all the more reason to believe she did not initiate or lead the revolt against Dunois' planned attack on St. Laurent before attacking the Tourelles during the battle for Orleans. More importantly, her actions support again the belief that Joan acted like a military professional.

Further, at Jargeau, Joan again displayed her military genius, tactical intuition, and military prudence. After Joan ordered a full-scale assault on Jargeau, Alencon, the commander appointed by Charles VII to lead the Loire River campaign, protested, "I told her it was premature to attack so quickly."[10] But Joan responded, "Have no fear....it is the right time when it pleases God."[11] Upon this reason, the French army attacked. Once again, Joan seized the "favorable moment," the "great art" as defined by Napoleon.[12]

On the night of June 14, 1429, after returning to Orleans, Joan requested the company of Alencon. According to Alencon's chronicler, Joan told him, "Tomorrow after dinner, I want to go to Meung. Have the company ready to leave at that hour."[13] For those, like Anatole France, who argue that Joan never commanded anything,[14] this simple anecdote shows who was really in charge. During the next few days both Meung and Beaugency were taken, clearing the Loire Valley and placing the English in a full retreat toward Paris.

At this point it is important to note that by attacking the Tourelles, the strongest point in the English defenses at Orleans, and by then capturing the extreme ends of the English fortifications along the Loire River, the French threatened the English center, Paris. It was a brilliant offensive strategy designed to secure the Bourges salient and free the king's passage to Reims, but it also leveraged the greater political goal of isolating

---

[10]Ibid., 277.

[11]Ibid.

[12]Philips, ed., "Military Maxims of Napoleon" in *Roots of Strategy*, 436.

[13]Parr, I, 160; Lang, 159; Bangs, 152; Richards, 167; Paine, I, *Joan of Arc*, 213; Paine, *The Girl in White Armor*, 115; Waldman, 147; Sackville-West, 194-95; Gies, *Joan of Arc*, 94; Richey, 73.

[14]In this instance, to avoid placing Joan in command, France, I, 362n, assigns this statement solely to an invention by the clerk writing the chronicle; whereas, in other instances while quoting this clerk no such disclaimers are made. See also Ronald Edward Zupko, "The Many Faces of Joan," in *Joan of Arc and the University*, Tallon, ed., 10.

Burgundy from England. These medieval minds were not barren of military strategy and of its need to support political strategy, as so often disparaged by medieval historians.

The English outposts all along the Loire River were finally consolidated and then joined by approximately 1,500 reinforcements from Paris, totaling 5,000.[15] Meanwhile Joan and the French, who received 1,000 reinforcements of their own, tallied approximately 6,000[16]—pretty much a fair fight. But the French, not content with simply occupying their repatriated towns and gloating over their victories, decided to pursue the English now retreating toward Paris. During this retreat, we have a perfect example of why discipline is so important to an army. In the modern military, a significant degree of soldier discipline is measured by observing noise, light, and litter discipline.[17] The historic Battle of Patay would be decided not by a quirk of fate, but by noise discipline.[18]

---

[15]France, I, 368; Paine, I, *Joan of Arc*, 217; Lynch, 161; Rutter, 196; Stolpe, 140; David-Darnac, 80; Gies, *The Knight in History*, 178; Maxwell, 94; De Vries, 111,116; Nicolle, *Orleans, 1429*, 76; Spoto, 87; Harrison, *Joan of Arc*, 177. Holinshed, "Chronicles of England, Scotland, and Ireland," in *Joan of Arc: Fact, Legend, and Literature*, Jewkes and Landfield, eds., 85, says 6,000 English; Smith, 72, says 5-6,000; Adams, 126, counts 4,000; while Burne, 252 says only 3,000. However, if 4,000 English survived Orleans and seven hundred English were lost at Jargeau, and if Fastolf arrived with the same 1,500 soldiers who fought the Battle of the Herrings, then the English force would have numbered closer to 5,000.

[16]Paine, I, *Joan of Arc*, 220; Burne, 256; Pernoud, *Joan of Arc*, 116; Gies, *The Knight in History*, 180; De Vries, 117; Nicolle, *Orleans, 1429*, 79. This number apparently reflects the departure of three thousand French militia after their victory at Jargeau, which conduct from the militia is not unreasonable after their pummeling by English archers, since their usual purpose for joining an army was to plunder and pillage, which purpose they achieved at Jargeau, whereas the smaller villages of Meung and Beaugency offered little of value. Lang, 164, and Paine, I, *Joan of Arc*, 223, report that Wavrin thought there were 12,000 French, while Monstrelet and Parr, I, 160, report 8,000 French at this time. Spoto, 87, says only 2,000.

[17]Noise and light can give away troop locations and actual numbers to the enemy, or they can be used to deceive the enemy as to these same things. Washington used large campfires to deceive the British while secretly retreating from Brooklyn Heights during the American Revolutionary War. Antal, 84. The Union General Rosecrans did the same thing to deceive the Confederates before crossing the Tennessee River and capturing Chattanooga without losing a man during the American Civil War. Tucker, 16. Litter can be used to track an enemy and can also reveal a lot about their logistics, training, and preparedness. It can also be used for deception: Rosecrans sawed off the ends of boards and threw them into the Tennessee River and had his troops pound on empty barrels to simulate boat building at a feigned river crossing site. Ibid.

[18]Shakespeare emphasized the need for noise discipline in *Henry V* the night before the Battle of Agincourt, insisting that it was a law of war: "In the name of Jesus Christ, speak lower…If you would but take the pains but to examine the wars of Pompey the Great, you shall find, I warrant you, that there is no tiddle-taddle nor pibble-pabble in Pompey's camp." Harrison, *Henry V*, Act IV, Scene I, lines 65,68-71, pp.756-57. Concerning George Washington's revolutionary

With the English archers hidden in ambush, awaiting the pursuing French cavalry, a stag, flushed by La Hire's advancing French scouts, suddenly bolted across an open field and directly toward the concealed English positions.[19] English troops reacted by abandoning their stealth, erupting in undisciplined whoops and cheers of delight, which easily alerted the forward scouts of La Hire's advancing cavalry. The results were thousands of lives lost. No wonder Henry V threatened the confiscation of horse and armor or the loss of an ear from anyone who made noise the night before the Battle of Agincourt.[20] When combat veterans of WW II were asked about the most common mistakes made by new soldiers, they answered "bunching up" and "talking loud or making noise at night."[21]

History tells us that La Hire sent word of this contact back to the French commanders[22] who asked Joan, "What should we do?"[23] Joan answered, "Have you good spurs!"[24] Astonished, the French commanders thought she meant retreat, which response unwittingly reveals their mindset, but Joan further explained that their spurs would be necessary to pursue the English.[25]

With these orders, La Hire's cavalry, revenge-soaked in memory, surprised the English with their own ambush and butchered the English soldiers at will. The rest of the English army, witnessing this onslaught, attempted to flee but were cut down in massive numbers. Estimates of

---

army, "Another specifically Prussian element which Steuben [Eric von Steuben had served under Frederick the Great] introduced was the relative silence with which every maneuver and action, from loading the rifle to advancing, was carried out." Koch, 185.

19 Lamartine, 133; Adams, 126; Lowell, 134; Oliphant, 105; France, I, 373; Lang, 165; Whetnall, 120; Bangs, 158; Wilmot-Buxton, 95; Richards, 174; Lynch, 165; Taylor, *Joan of Arc*, 161; Ince, 34; Ferguson, 65; Lamond, 95; Paine, I, *Joan of Arc*, 226; Paine, *The Girl in White Armor*, 123; Endore, 124; Waldman, 154; Sackville-West, 200; Rutter, 197; Buchan, 132; Burne, 258; Leary, 67; Brown, 635; Pernoud, *Joan of Arc*, 118; David-Darnac, 83; Beevers, 96; Gies, *Joan of Arc*, 99; Wooster, 116; Pernoud and Clin, 61; De Vries, 119; Richey, 71; Taylor, *The Virgin Warrior*, 84; Goldstone, 142; Nicolle, *Orleans, 1429*, 80,82; Harrison, *Joan of Arc*, 177; Castor, 117.

20 Haggard, 120; Neillands, 214. See also Burne, 77.

21 McManus, 313.

22 Bray, 151; Bangs, 156; Richards, 174; Fabre, 187. This information comes from Jean de Wavrin (see Pernoud, *Joan of Arc*, 118) who was engaged in this battle on the side of the English and therefore could not have known what the French messengers were doing. Dunois gives testimony (Murray, 238) suggesting that Joan's orders were in response to the English army departing Meung rather than to any messages from La Hire, although the timing within his testimony is vague. If such was the case, then La Hire was simply following orders to "have good spurs."

23 Bray, ibid.; Lowell, 133; Wilmot-Buxton, 95; Richards, 172.

24 Murray, 238.

25 Ibid.

English casualties reach as high as 4,000 dead and as many as five hundred prisoners,[26] while the French suffered just one or two casualties.[27] The Battle of Patay was an overwhelming French triumph. Mark Twain and Andrew Lang recorded it as "the Moscow of the English power in France,"[28] referring to Napoleon's infamous defeat in 1812. La Hire himself captured Lord Thomas Scales. Lord John Talbot, too, was made prisoner.[29]

"Have you good spurs" is a brilliant metaphorical transmission of Joan's combat vision, her overall commander's intent. In accordance with modern U.S. Army leadership doctrine, "Leaders in command positions use commander's intent to convey purpose. Commander's intent is a *concise expression* of the purpose of the operation and the desired end state....This is important for situations when unanticipated opportunities arise or the original solution no longer applies."[30] *Give chase to the English* was the purpose of Joan's statement as she further explained, but more importantly, it reinforced the aggressive nature of the combat operations she expected from these French commanders. Simple, direct, precise, and unequivocal, "Have you good spurs" again challenges Madison Avenue with the magnificence of its eloquence, and makes clear and direct the precise brand of Joan's desired military product—unremitting aggressiveness. After all, these exact words were still remembered more than twenty years later.

"Commander's intent is a concise expression of the purpose of the operation and the desired end state."

Having destroyed the English army, Joan departed Gien for Riems to crown the king on June 27, 1429, riding with the French army's advance guard, once more up front with La Hire. Two days later Charles followed and embarked on his first military campaign in years with a large staff of religious and political advisers in tow.[31] Although most of the English

[26]Murray, 238; Sackville-West, 200-01, gives an excellent summary of the conflicting casualty figures cited from various sources which range from 1,800 to 4,000 dead and from two hundred to five hundred prisoners. Meanwhile, Clausewitz, 669, again reminds us, "Casualty reports on either side are never accurate, seldom truthful, and in most cases deliberately falsified."

[27]Paine, I, *Joan of Arc*, 229; Gies, *Joan of Arc*, 99. Pernoud, *Joan of Arc*, 119, says three.

[28]Twain, 443; Lang, 65; Whetnall, 83.

[29]Sowers, trans., *Chronicles of St. Denis*, x; Adams, 128; Lang, 166; Paine, I, *Joan of Arc*, 229; Castor, 118. See also Pierre Champion in Barrett, 455.

[30]Army Leadership, 6-4.

[31]Lowell, 144; France, I, 403; Lang, 172; Waldman, 166; Sackville-West, 204; Stolpe, 153.

and Burgundian held cities capitulated in the face of this advancing French army, which ballooned to 12,000 as ever more volunteers flooded their ranks, when they reached the city of Troyes their march to Reims was halted. Troyes, the scene of the Treaty of Troyes and Henry V's political triumph, decided to resist. Negotiations ensued for three days before Charles VII held counsel with his commanders on what to do.

Joan arrived late for this meeting, listened to the council's verdict, which advised retreat, and asked the king if he would believe what she had to say. Charles agreed, hesitantly, at which point Joan promised to take the city within three days.[32] Even the French officers were stunned by Joan's rashness. Joan's only response was to turn abruptly to the king and declare, "Doubt not but that *tomorrow* you shall be master of this city."[33] Without further debate, Joan defiantly mounted her horse and ordered preparations for a French attack.

Early the next morning an immediate deployment was ordered by Joan and the French army quickly encircled half the city. Artillery rumbled forward and faggots were assembled to fill in the city moats at selected crossing sites.[34] One French commander described events this way: "Then Jeanne, putting herself at the head of the army, had the tents placed right against the trenches of the town, and executed many marvelous maneuvers which had not been thought of by two or three accomplished Generals working together."[35] Clearly, once again, Joan was in command.

With daybreak, the citizens of Troyes heard the loud clamor of armor as ladders broke the dawn skyline and French captains emerged from the early morning mists in all their combat plumage. No doubt Joan's banner unfurled before the trembling sight of stationed observers along the walls of Troyes. Joan gave the order to attack and faggots were thrown in to fill

---

[32]Murray, 239; Michelet, *Joan of Arc*, 40; Bray, 159; O'Hagan, 58; Lowell, 150; France, I, 427; Richards, 189; Taylor, *Joan of Arc*, 165; Ferguson, 69; Lamond, 103; Monahan, 133; Endore, 137; D'Orliac, 190; Rutter, 210; Winwar, 127; Fabre, 196; Stolpe, 158; Pernoud, *Joan of Arc*, 122; Guerin and White, 108; Smith, 76; Gies, *Joan of Arc*, 108; Maxwell, 98; Pernoud and Clin, 63; De Vries, 132; Taylor, *The Virgin Warrior*, 86; Harrison, *Joan of Arc*, 183. Sowers, trans., *Chronicles of St. Denis*, xiii; Rankin and Quintal, trans., *The First Biography of Joan of Arc*, 36; Parr, I, 182; Adams, 135; Paine, I, *Joan of Arc*, 256; Paine, *The Girl in White Armor*, 134; Waldman, 171; Sackville-West, 207; and David-Darnac, 95 state "two days," while Ireland, trans., II, *Journal of the Siege of Orleans*, 98, Haggard, 193, Lynch, 180, and Castor, 124, state "two or three days."

[33]Michelet, *Joan of Arc*, 40; Parr, I, 182; Bray, ibid; Tuckey, 74; O'Hagan, 59; Lowell, 150; Oliphant, 116; Haggard, 293; Madison, 271; Ferguson, 69; Lamond, 104; Monahan, 133; Fabre, 196; Pernoud, *The Retrial of Joan of Arc*, 110; David-Darnac, 95; Wooster, 120.

[34]Sowers, trans., *Chronicles of St. Denis*, xiii; Parr, I, 182; Pernoud, *The Retrial of Joan of Arc*, 110.

[35]Murray, 239-40.

up the city moats.[36] Suddenly, and without any fighting, the gates swung open and the city surrendered.

The French army finally arrived at Reims on July 16, a Saturday,[37] and the next day Charles VII was officially crowned the king of France. As a reward to Joan, Domremy citizens remained tax free until the eighteenth century with all accounts reading only: "Nothing—the Maid."[38] But to Joan's utter desolation, amidst all the glory of her success, the world of her special voices suddenly...went silent.[39]

Meanwhile, the English quickly rebuilt their army. By fortunate circumstance, 3,500 fresh crusading troops[40] were conveniently disembarking from the ports of Pas de Calais in mid-July just when the duke of Bedford needed them. He immediately sent a messenger north who convinced them to divert their army south to Paris instead of east to Bohemia.[41] Meanwhile, Bedford strengthened his capital defenses. The protective earthworks in front of the six city gates were reinforced,[42] moats were deepened,[43] artillery was brought in,[44] munitions were increased,[45] provisions stockpiled,[46] and all possible stores and equipment outside the city were either moved inside or scorched.[47] Like Orleans before, Paris made ready for siege warfare. The Bohemian crusaders arrived in Paris on July 25.[48]

Joan had accomplished the mission given by her voices, so why not go home? The most likely reason for remaining under arms came not from her

---

[36]Paine, I, *Joan of Arc*, 360n; Pernoud, *Joan of Arc*, 123; Pernoud, *The Retrial of Joan of Arc*, 110.
[37]Tuckey, 78; Lowell, 154; Bangs, 187; Lynch, 183; Endore, 141.
[38]Bray, 169; Adams, 148; Lowell, 156; Bangs, 192; Stevens, 148; Richards, 201; Lamond, 111; Paine, I, *Joan of Arc*, 272; Paine, *The Girl in White Armor*, 143; Endore, 144; Sackville-West, 215; Winwar, 137; Warner, *Joan of Arc*, 168; Spoto, 84.
[39]See also Adams, 19, Whetnall, 131; Stevens, 151,159, Buchan, 140, Fabre, 219, and Vale, 58.
[40]Stolpe, 159; Leary, 71; Pernoud, *Joan of Arc*, 129; Guillemin, 126; Gies, *Joan of Arc*, 118. Adams, 131, counts 250 lances (750-1,250) plus 2,500 bowman (3,250-3,750). Castor, 134, therefore scores 2,500 archers and 250 men; Lang, 168, counts 3,350 crusaders; Ireland, trans., II, *Journal of the Siege of Orleans*, 107; Waldman, 181, and Winwar, 140, count 4,000. Parr, I, 199 Lynch, 192, Endore, 147; and Fabre, 209, say 5,000, and David-Darnac, 105, reports 6,000.
[41]France, II, 441-42; Warner, *Joan of Arc*, 176-77; Gies, *Joan of Arc*, 117; Pernoud and Clin, 71.
[42]Pernoud, *Joan of Arc*, 134; De Vries, 136.
[43]Pernoud, ibid.; David-Darnac, 109; De Vries, ibid.; Goldstone, 151.
[44]Pernoud, ibid., 120; De Vries, ibid.; Goldstone, ibid.
[45]Bray, 184; Gower, 86-7. One stone-cutter alone constructed 1,176 cannonballs. Winwar, 142; Lucie-Smith, 176; De Vries, ibid.; Goldstone, ibid.; Harrison, *Joan of Arc*, 205
[46]Paine, I, *Joan of Arc*, 318.
[47]Gower, 86-7; Pernoud, *Joan of Arc*, 134; Lucie-Smith, 175-76; De Vries, 136.
[48]Michelet, *Joan of Arc*, 47; Gower, 83; Lowell, 164; Lang, 188; Paine, I, *Joan of Arc*, 289; Waldman, 181; Stolpe, 159; Pernoud, *Joan of Arc*, 129; David-Darnac, 105; Guillemin, 126; Pernoud and Clin, 72.

voices or from her faith or from Charles VII. Her real dilemma, absent her voices and family, was her soldier soul which forced her to remain loyal to these citizen militia and professional regulars who had so passionately volunteered to ride and fight beside her, who had believed and trusted in her, and who had died or would die so readily for her. If there was going to be further combat, Joan could not abandon these faithful Frenchmen. Agnes Royden recognized this loyalty with the wonderful statement, "There is nothing lovelier in history than the loyalty of the Maid to her soldiers and the loyalty of her soldiers to the Maid."[49] And that mutual soldier bond has stretched into modern military history. For example, at Selles sur Cher, during WW I, the Catholic soldiers of the U.S. Army donated a statue of Joan to the local church in memory of their stay in that town from January 1918 to June 1919.[50] These universal soldier loyalties to Joan are evident everywhere in France and resplendent in their nature, just as Joan's loyalty to her soldiers was equally magnificent.

Joan was a true military commander who could not and would not forsake her soldiers, whose powerful camaraderie now prevailed, even above her loyalties to family, voices, and Charles. Whether Joan understood her dilemma or not, she went against everything she knew because she had no choice: "the saint had become a captain."[51] As Janice Tuckey so marvelously observed, "She had given a King to France, now she had to give France to her King"[52]—thus she would continue personally to lead her soldiers...for them and for the sake of France.

"the saint had become a captain."

Loyal Frenchmen had relied on Joan and died for her, and their blood had bonded to her soul. And she felt this bond by understanding that every soldier's life was someone's father, son...or brother: "For he today that sheds his blood with me/Shall be my brother."[53] Thus, according to her personal priest, Joan's only wish, her only wish if she were to die, was for the king to build a chapel "where the souls of those who died in defense of the kingdom might be prayed for."[54] No thoughts of mother, father, sister, friends, kings or queens. Joan wanted a chapel to honor the spirit of her comrades and to

[49]Royden, 44.
[50]Rutter, 74.
[51]Adams, 143
[52]Tuckey, 81.
[53]Harrison, *Henry V*, Act IV, Scene iii, lines 61-2, p.761.
[54]Murray, 290; Michelet, *Joan of Arc*, 54; Parr, I, 206; France, I, 342; Adams, 156; France, I, 342.

immortalize their sacrifice. The battlefield bond had finally possessed her, and because of that bond she remained with her army and with its soldiers, soldiers she thought of first above all else—her only wish being to pray for them. Even if she had wanted to, Joan could not abandon them...this her army. With her victory at Orleans, and with Charles' coronation at Reims, Joan of Arc had become "the Maid of France"—and she could not escape it.

Joan entered St. Denis, just five miles outside Paris, on August 26, 1429, with a relatively small force of French professionals. They found the city virtually abandoned.[55] From St. Denis Joan and her soldiers conducted numerous reconnaissance forays against the Parisian walls, probing several times daily.[56] Finally, on September 7, Charles VII joined his army and the order was given to begin the battle for Paris.

From the village of La Chapelle,[57] located between St. Denis and Paris, the French army attacked on 8 September. Joan quickly galloped across the first moat, which turned out to be dry, only to be trapped in front of the second wider moat full of deep water. There Joan stopped, encouraging her soldiers to fill in the ditch, but they lacked sufficient faggots.[58] The attack stalled as French soldiers departed to find more wood. Meanwhile Joan sat on her horse before the walls of Paris demanding the English surrender. With the arrival of nightfall,[59] Joan, still in-between moats, was hit in the thigh by an arrow, even as other bow-shots impaled her horse, toppling it down a slope trapping her beneath its heavy flesh.[60] An anonymous soldier, carrying her banner alongside, was also instantly struck, first in the foot, and then between the eyes when he raised his visor to examine the damage from his first wound.[61]

---

[55]Monstrelet, 562; France, II, 49; Lang, 199; Lynch, 205; Lucie-Smith, 174; De Vries, 147.

[56]Parr, I, 221; Bray, 181; France, II, 51; Lang, 199; Bangs, 208; Lynch, 206; Paine, I, *Joan of Arc*, 310; Pernoud, *Joan of Arc*, 136; De Vries, 149.

[57]Parr, I, 223; Lowell, 174; Waldman, 189.

[58]Sowers, trans., *Chronicles of St. Denis*, xx; "The Chronicles of France" in *Joan of Arc: La Pucelle*, Taylor, trans., 253; Tuckey, 90-1; Gower, 89; Waldman, 191.

[59]According to Percival de Cagny, *"After the sun had set,* the Maid was hit by a crossbow bolt in her thigh (italics added)." Lowell, 176; Pernoud, *Joan of Arc*, 137; Pernoud and Clin, 77. The *Chronicles of St. Denis* (Sowers, trans., xx) places this wound earlier in the day.

[60]"Bourgeois of Paris" in *Joan of Arc: La Pucelle*, Taylor, trans., 230; Ireland, trans., II, *Journal of the Siege of Orleans*, 116,223n,225n; Michelet, *Joan of Arc*, 48; Bray, 186; Stevens, 164; Lynch, 208; D'Orliac, 206; Sackville-West, 222; Rutter, 223; Buchan, 153; Smith, 85; Grayeff, 66; Gies, *Joan of Arc*, 125; Taylor, *The Virgin Warrior*, 100.

[61]Taylor, trans., *Joan of Arc: La Pucelle*, ibid.; Parr, I, 225-26; Bray, ibid.; Lowell, 176; Oliphant, 165; Stevens, 162; Lynch, ibid.; D'Orliac, ibid; Sackville-West, ibid.; Rutter, 223-24; Winwar, 145; Leary, 80; Guerin and White, 129; Gies, ibid.; De Vries, 152; Harrison, *Joan of Arc*, 206-07; Castor, 142.

The French rescued Joan and moved her against her will to a safe location.[62] Nevertheless, she refused to leave the field and continued to exhort her soldiers to fill-in the moats and continue their attack.[63] But the effect was not the same as when she had reappeared so dramatically at Orleans as if resurrected from the dead. As the hours grew late, she was called to council, but Joan refused, remaining in contest.[64] Finally, near ten or eleven o'clock that night, the battle was halted.[65] Joan complained bitterly that the city might have been taken.[66]

Unlike the battles for St. Loup, the Tourelles, or during the Loire River campaign, the French attack on Paris did not enjoy a 3:1 or greater tactical advantage. For every French soldier scaling the walls, there was a defender waiting for him plus another in reserve. Further, the assault points were few and narrowly defined, channeled in a way that could easily be anticipated by the defenders. French soldiers could not suddenly appear everywhere at once all along the walls, which increased the concentration of defenders opposing them and reduced the combat ratios for the French. And unlike at the Tourelles, nightfall did not redefine the situation. The Parisians could still identify and defend against those few previously defined assault routes available to the French. In that single day at

---

[62]Parr, I, 227; Tuckey, 92; Lowell, 176; Bangs, 212; Stevens, 163; Lynch, 208; Paine, I, *Joan of Arc*, 314; Paine, *The Girl in White Armor*, 161; D'Orliac, 206; Sackville-West, 222; Fuller, *Decisive Battles*, 492; Pernoud, *Joan of Arc*, 137; Guillemin, 139; Gies, *Joan of Arc*, 125; Pernoud and Clin, 77; Taylor, *The Virgin Warrior*, 100.

[63]Lowell, ibid.; Lang, 206-07; Paine, I, ibid. This circumstance lends credence to Dunois' testimony describing Joan's reaction after her wound at the Tourelles—that she remained on the battlefield afterward—but he may have confused the two battles. According to Joan, this wound healed in five days, whereas her wound at Orleans took twenty days to heal (Murray, 14), suggesting the wound at the Tourelles was four times more severe.

[64]"The Chronicles of France" in *Joan of Arc: La Pucelle*, Taylor, trans., 253; Bangs, 212.

[65]Michelet, *Joan of Arc*, 49; Tuckey, 91-2; Adams, 155; France, II, 69; Paine, I, *Joan of Arc*, 318; Endore, 165; Winwar, 146; Stolpe, 176; De Vries, 153; Taylor, trans., *Joan of Arc: La Pucelle*, 125. Not only does the "Bourgeois de Paris" in *Joan of Arc: La Pucelle*, Taylor, trans., 230, exaggerate the French strength at 12,000 and embellish the events surrounding Joan's wound, he also ends the attack at four p.m. Despite these discrepancies, which seem to have come from the bottom of a wine cellar where the medieval author likely hid throughout the proceedings, some biographers elect to invoke his version of the Paris battle. For example see Pernoud, *Joan of Arc*, 135, Lucie-Smith, 179, De Vries, 152, and Harrison, *Joan of Arc*, 207. He obviously wrote his account after the fact and even inserted alleged English opinions about Joan and the other French commanders (see "The Bourgeois of Paris" in Taylor, ibid.), which he probably learned through hearsay.

[66]Parr, I, 227; Bray, 187; Oliphant, 166-67; Richards, 211; Paine, Ibid., 314; Pernoud, *Joan of Arc*, 136.

Paris, the French suffered 1,500 casualties, including five hundred dead.[67] Charles VII ordered the battle halted and the army returned to Gien where it disbanded.

In late October or early November, 1429, a small French army, exact size unknown,[68] set out from Bourges for St. Pierre-le-Moutier along the Loire River. Again Joan was not the designated commander when the initial assault was ordered. Eventually the attack bogged down and all but a few soldiers pulled back. Joan however stayed on the field and ordered the remaining soldiers to continue filling in the moat and to attack again. These soldiers did what they were told, the moat was filled, the attack was continued, and the town taken without further resistance.[69]

On November 24, the French laid siege to La Charite, a heavily fortified town with walls and towers, a central castle complex, and a single narrow bridge across the nearby river. The town was bombarded daily and at least one assault attempted, but winter arrived early that year and proved brutally cold.[70] The siege was finally abandoned, marking Joan's second failure. Nevertheless, on December 29, 1429, just a little more than ten months after her departure from Vaucouleurs, Charles ennobled the entire d'Arc family through its entire succession, both the male *and female* line, an unprecedented act of royal charity.[71]

Joan sat out the winter months with the royal entourage and in April 1430 began again her attacks on English positions. While in a moat outside Melun her now intermittent voices returned and their message was ominous. St. Catherine and St. Margaret warned that she would be captured.[72]

---

[67]Paine, I, *Joan of Arc*, 317; Leary, 80; Gies, *Joan of Arc*, 126; Taylor, *The Virgin Warrior*, 101; Harrison, *Joan of Arc*, 206. See also Michelet, *Joan of Arc*, 49, and Adams, 155.

[68]Leary, 82, counts the army at 2,000, but without footnotes.

[69]Ibid.

[70]Smith, 90; Gies, *Joan of Arc*, 135.

[71]Evans, 151; Bartlett, 170; Parr, I, 250-53 (contains the entire decree); Bray, 169,196; Tuckey, 97; Adams, 159; Lowell, 194-95; Julleville, 79; Lang, 221; Whetnall, 142-43; Bangs, 229; James, 231; Wilmot-Buxton, 121-22; Stevens, 174; Richards, 220; Lynch, 219; Monahan, 167-68; Paine, II, *Joan of Arc*, 17; Paine, *The Girl in White Armor*, 171; Endore, 179-80; D'Orliac, 215; Waldman, 202; Sackville-West, 236; Rutter, 239; Buchan, 159; Winwar, 138; Fabre, 232; Stolpe, 184; Guerin and White, 136; Pernoud, *Joan of Arc*, 145-46; Guillemin, 147; Raknem, 14; Smith, 91; Lucie-Smith, 193; Grayeff, 70; Warner, *Joan of Arc*, 165; Gies, *Joan of Arc*, 134; Marot, 66; Maxwell, 121; Pernoud and Clin, 81; De Vries, 165; Taylor, *Joan of Arc: La Pucelle*, 176n; Spoto, 109; Taylor, *The Virgin Warrior*, 107,208; Goldstone, 156; Harrison, *Joan of Arc*, 212; Castor, 150. However, Sermoise, 199, believed this document was "patently faked," and others suspect it was a forgery, first surfacing around 1596. Warner, 195.

[72]Murray, 57.

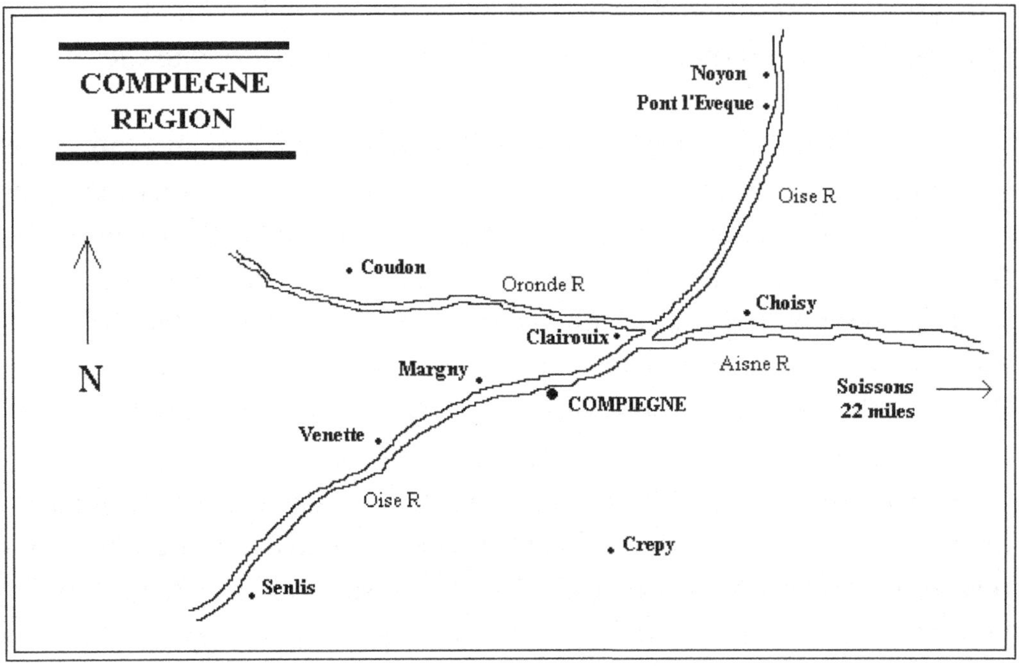

Afterwards these saints resumed their daily visitations and with each arrival these voices, who normally carried only words of encouragement, repeated their message of impending doom.

In early May, 2,000 English and 4,000 Burgundians[73] neatly arrayed themselves all along the Oise River with the city of Compiegne at their center and directly in their sights. Joan rode to Compiegne to join in the fight against them. With a small force of 500, she attacked directly into the center of the English and Burgundian encampments and was soon encircled by enemy troops. Seeing themselves trapped, the French began a frantic withdrawal in the direction of the Compiegne bridge, but Joan refused to follow and continued to fight. Finally, some French soldiers grabbed the reins of her horse and forced her to ride for the city while she used her sword to fend off any pursuers.[74]

Nonetheless, Joan and her small company were stopped and surrounded by a bevy of Burgundian soldiers who yelled at her to surrender, each imagining the riches and rewards for her capture. But Joan refused and struck out with her sword, keeping her pursuers at bay. Finally an archer named Lyonnel, in service to Guillaume, the Bastard of Wandonne, snuck

[73]France, II, 140; Endore, 195.

[74]Murray, 333; Oliphant, 194; France, II, 151; Wilmot-Buxton, 130; Rutter, 244; Stolpe, 190.

behind and grabbed Joan by her cloak and unceremoniously dumped her from her horse.[75]

In researching the events around Compiegne, it remained a personal mystery for many years as to why Joan would so foolishly attack into the teeth of 6,000 soldiers with just 500 men. As it turned out, the answer to this mystery was elusive not because of any foolishness, but rather because of the level of sophistication Joan had reached in her military thinking.

Most likely the location for any English or Burgundian commander would have been near the center of his command. Recall that when Joan arrived at Orleans, she immediately wanted to go after Talbot and when Sir William Glasdale was killed in the Tourelles, Talbot immediately abandoned the field. Later, at the Battle of Patay after Talbot was captured, the English commander, Sir John Fastolf, also abandoned the fight. Joan may have expected the same results if she captured or killed either John II of Luxembourg or the duke of Burgundy—the rest of the English would run.

Luxembourg, however, was actually headquartered at Clairoix, and the duke of Burgundy at Noyon, away from the center of their commands where Joan chose to attack. Nevertheless, Pope Pius II believed that Joan had intelligence information placing the duke of Burgundy on frequent hunting trips[76] along the rivers, which would explain the time of day for this excursion (5 p.m.), and the reason for her ill-advised excursion.[77] Whatever the case, Joan's *coup de main* to capture either Luxembourg or the duke seemed her best chance for ending the siege quickly and is a more reasonable explanation for why she persisted so adamantly with so few men—to locate and draw Luxembourg or the duke of Burgundy personally onto the field of battle. This decapitation strategy was a common practice exercised by both Alexander the Great and the duke of Wellington.[78] In the second Persian Gulf War, America also attempted a quick victory by attempting to

---

[75]Murray, 335; Bartlett, 180; Parr, I, 273; Bray, 209; Lamartine, 134; Tuckey, 107; Lowell, 219; France, II, 152; Lang, 240; James, 279; Stevens, 188; Lamond, 139; Monahan, 176; D'Orliac, 224; Rutter, 244; Winwar, 159-60; Stolpe, 190; Leary, 86; Lightbody, 70; Pernoud, *Joan of Arc*, 151; Sermoise, 99; Lucie-Smith, 205; Warner, *Joan of Arc*, 75; Gies, *Joan of Arc*, 141; Beevers, 117; Maxwell, 127; Pernoud and Clin, 88; Spoto, 112; Harrison, *Joan of Arc*, 227.

[76]Late evening, after 5 p.m., is always a good time to begin hunting.

[77]Gragg, trans., 207. Stolpe, 189, also mentions the possibility that Joan wanted "to capture the commander-in-chief" of the allied forces, but he uses no footnotes. He then goes on to suggest that she was simply conducting a reconnaissance while at the same time attempting to take the village of Margny, which makes no sense; reconnaissance elements generally try to remain undetected and unengaged.

[78]Keegan, *The Mask of Command*, 148.

decapitate the head of the Iraqi army, Saddam Hussein, with their initial air strikes.[79]

Custer died, Henry V escaped with victory, but Joan was captured when heavily outnumbered. The fortunes of war are rarely predictable.

News of Joan's capture reached Paris on May 25, and the very next day, in the name of the Inquisitor of France, Martin Belormi[80] (some names otherwise obscure in history should be remembered), on behalf of the University of Paris—scholars, intellectuals, masters, doctors, and wise men all—immediately wrote to the duke of Burgundy demanding that Joan be handed over for investigation into crimes of heresy and witchcraft.[81]

Joan attempted to escape her imprisonment twice: once by prying up some flooring in her prison cell; and again when she flung herself off a sixty foot high castle tower, knocking herself unconscious. Finally, in November 1430, she was sold to the English and put on trial in the French city of Rouen, the English not trusting the security around Paris.

On January 3, 1431, King Henry VI officially released Joan to a church tribunal. On February 21, 1431, the day after the Pope died in Rome,[82] Joan was brought into court on unspecified charges. To describe the trial itself would require another book. Suffice it to say, the outcome was never in doubt and Joan was found guilty. She was burned at the stake on May 30, 1431.

More than twenty years later, the Catholic Church convened what has been called a nullification trial. The church found her innocent of all charges in 1456. She was canonized a Catholic saint in 1920.

---

[79]Powell, 114.

[80]Parr, II, 2. David-Darnac, 133, spells the name "Billorme," Stolpe, 193, spells the name "Billoray," and Hobbins, 34, spells it "Billorin."

[81]Murray, 336-37; Rankin and Quintal, trans., *The First Biography of Joan of Arc*, 44-6; Parr, ibid.; Tuckey, 108-09; Gower, 102; O'Hagan, 70; Lowell, 227; Oliphant, 204-07; Julleville, 91; France, II, 156; Lang, 243; Bangs, 253; James, 281; Wilmot-Buxton, 135; Stevens, 193; Lynch, 241-42; Denis, 95; Paine, II, *Joan of Arc*, 64; Lamond, 144-45; Paine, *The Girl in White Armor*, 189-90; Barrett, 7-9; Waldman, 220; Sackville-West, 259; Fabre, 251; Stolpe, ibid.; Pernoud, *Joan of Arc*, 157; Guillemin, 170; Smith, 104; Lucie-Smith, 207; Cristiani, 118; Warner, *Joan of Arc*, 96; Gies, *Joan of Arc*, 143,259; Wooster, 136; Pernoud and Clin, 91; De Vries, 183; Fraioli, 159; Spoto, 112; Vale, 59; Goldstone, 166; Castor, 161.

[82]Wyndham, 22; France, II, 250; Rousseil, 177; Brown, 480, Pernoud, *Joan of Arc*, 237; Pernoud and Clin, 126. Lynch, 275, says 20 February.

Appendix A

THE ST. LOUP RESCUE

The scenario for Joan's participation in the battle for St. Loup will not be found in any historical footnotes. What is more, other accounts of her heroics in this battle remain equally unsupported, although various authors over time have contrived and heavily embellished her actions against this outpost. While the conclusions given here are based on the barest of evidence, they are more likely right than wrong as to what actually happened. Here are the reasons.

In a letter from Charles VII to the cities of France, he described the St. Loup battle with this statement: "...although the English who were in the other bastion [St. Pouair?] had come out in battle seeming to offer combat, nevertheless when they saw our people come to meet them, they turned about hastily without daring to wait for them,"[1] but he failed to name the French commander who led these forces. Nonetheless, *The Journal of the Siege of Orleans* places Saint Severe (also called Boussac[2]) and Graville in command.[3]

Michelet in 1841 was the first to place Joan at the head of this relief column,[4] without any supporting footnotes. Subsequently, Stanhope in 1853 named Boussac[5] as the only commander. Harriet Parr in 1866 placed Dunois at the head of this force.[6] Alphonse De Lamartine prior to 1880 again placed Joan in command,[7] while Janet Tuckey in that same year

[1]Paine, I, *Joan of Arc*, 347n.
[2]See Pierre Champion in Barrett, 403.
[3]Ireland, trans., II, *Journal of the Siege of Orleans*, 67.
[4]Michelet, *Joan of Arc*, 30-1.
[5]Stanhope, 42.
[6]Parr, I, 117.
[7]Lamartine, 131.

failed to name a leader. Lord Ronald Gower in 1893 retained Boussac and Graville as joint commanders.[8] Anatole France in 1908 again dropped Graville and mentions only Boussac.[9] Andrew Lang in 1909 ignored Anatole France and failed to name a commander.[10] However, Mary Bangs in 1910,[11] Denis Lynch in 1919,[12] Michael Monahan in 1928,[13] Jehanne D'Orliac in 1934,[14] and Milton Waldman in 1935[15] all followed France's assignment of Boussac as the only commander. Meanwhile Endore in 1931 returned both Graville and Boussac as joint commanders,[16] while Sackville-West in 1936 refused to name a leader. In 1954, Fabre continued with Graville and Boussac as the leaders and added Coulonces.[17] Gies, in 1959, simply has Joan riding out alone with a few companions.[18] David-Darnac, in 1965, returned the command to Boussac only.[19] Finally, Nicolle, in 2001, repeats Fabre with Boussac, Graville, and Coulonces as the commanders. (Leonard Wolf, in 1980, unexplainably places de Rais and La Hire in command of this French counterattack.)[20] Thus based solely on *The Journal of the Siege of Orleans*, Boussac and sometimes Graville are generally accepted as the primary commanders for this assault. But these accounts by Michelet, Stanhope, Parr, LaMartine, Gower, France, Lang, Bangs, Lynch, Monahan, D'Orliac, Waldman, Endore, Fabre, David-Darnac, Wolf, and Nicolle are all without supporting footnotes[21] (many authors simply omit Talbot's counterattack).

Similarly, Lang, (1909), Bangs (1910), Lynch (1919), Paine, I, *Joan of Arc* (1925), Monahan (1928), Endore (1931), Waldman (1935), Sackville-West (1936), Purcell (1952), Scott (1974), and Nicolle (2001), all claim that an anonymous force of "six hundred" swept forth from the gates of Orleans to intercept Talbot. However all of them neglect any footnotes or references concerning these entries, although these six hundred can be found

[8]Gower, 54.
[9]France, I, 288.
[10]Lang, 125.
[11]Bangs, 110.
[12]Lynch, 147.
[13]Monahan, 77.
[14]D'Orliac, 157.
[15]Waldman, 111.
[16]Endore, 97.
[17]Fabre, 151.
[18]Gies, Joan of Arc, 75.
[19]David-Darnac, 65.
[20]Wolf, 54.
[21]France, I, 289n, has a footnote, but it does not support his narrative.

in *The Journal of the Siege of Orleans*[22] which originally named Graville and Severe (Boussac) as the commanders.

All of this conflicting and unsupported evidence from a single source forced yet another search through the nullification trial testimonies, whereupon was found a critically important tidbit of relevant information, which not surprisingly comes from Louis de Contes who was with Joan at the eastern gate. He states, "The English were *preparing to resist* when Jeanne advanced against them in all haste" (italics added).[23] This testimony is important because of course the English were not *preparing to resist*, they were *already resisting* at St. Loup. In fact French wounded were arriving at the gates of Orleans. So this *preparing to resist* could not have been in reference to St. Loup but only in reference to sightings of Talbot's relief force which must have been on the march at the time.

St. Loup was two miles beyond the eastern gate. Two good men burdened with a stretcher (usually a converted shield) could travel that distance in something close to thirty minutes. Allowing time for troop engagement, recovery of wounded from the battlefield, and some treatment of their wounds, Joan must have been as much as an hour late for this three hour long[24] attack (assuming these were early French casualties). During this hour the militia must have waited for Joan or continued to assemble at the eastern gate. Thus, if de Contes is correct and the English were *preparing to resist*, he must have been able to view Talbot's counterattack responding from the nearer location of St. Pouair, where Talbot had stationed his reserve reaction force, to make sense of the timeline available. Consequently, Talbot must have selected a route of march that took him near enough to the eastern gate to allow Orleans sentries and de Contes to observe these English forces.

Viewing detailed maps of the Orleans area at that time (one of the best is in Leary, between pages 14 and 15); there were no easy routes from St. Pouair to St. Loup. In fact all roads cut vertically north-south and were perpendicular to his path (except for the road leading from the eastern gate of Orleans), which would have forced Talbot to travel cross country slowing

[22]Ireland, trans., II, *Journal of the Siege of Orleans*, 67.

[23]Murray, 262. J. M. Cohen in Pernoud, *The Retrial of Joan of Arc*, 177, translates this passage as the "English were preparing their defense," which again shows the dangers between interpretive versus literal translations. Obviously, English defenses were already prepared and in place at this time.

[24]Ireland, trans., II, *Journal of the Siege of Orleans*, 67; Parr, I, 117. France, I, 289; Lynch, 147; Monahan, 76; Nicolle, *Orleans, 1429*, 65; Harrison, *Joan of Arc*, 149; Castor, 108.

The route to St. Loup would have been from the traffic circle (center far left) to the last intersection (upper far right). (Image@2016 Google Earth)

his movement, whereas Joan and her forces had a straight and easy path to intercept him as he neared St. Loup. Finally, in further consideration of Talbot's delay, he likely sent out scouts first to report on the situation before deciding to commit a modest reaction force against what initially appeared to be an exposed French attack.

Moreover, Joan's militia were Orleans' based, as were La Hire's professionals, which more logically supports their inclusion in any prior planning for a prepared attack against St. Loup, as opposed to an ad hoc gathering by Boussac and Graville. Boussac and Graville had just arrived with Dunois that morning from Blois[25] and their troops were likely busy quartering themselves or otherwise arranging military affairs and thereby spread throughout the city. To insert them now into this midday battle requires some rather lofty assumptions. First, either Boussac and Graville had to be kept lying-in-wait at the Paris (northern) gate (France's version of their departure point to intercept Talbot[26]), or they had to have enough

[25]See Murray, 267, testimony of Simon Baucroix; Parr, I, 114; France, I, 283. Boussac had left the army shortly after the Battle of the Herrings because his brother-in-law had died in that battle and he needed to settle the estate and other affairs. Waldman, 100. However, there is sufficient time for him to return to the army and escort the convoy to Orleans.

[26]France, p.288.

time to gather these forces on an emergency basis. Realizing this choice, France elected to begin Talbot's counterattack from St. Laurent,[27] on the west side of the city, furthest from St. Loup, allowing watch guards from the Orleans' towers to spot Talbot's consolidation and thereby report his subsequent eastern movement in sufficient time for Boussac and Graville to assemble their troops by means of using church bells as alarms,[28] which bells, one would think, would have awoken Joan. Burne follows France in beginning Talbot's attack at St. Laurent.[29] However, *The Journal of the Siege of Orleans* specifically identifies St. Pouair as the departure point for Talbot's troops. Later, *The First Biography of Joan of Arc* (ca 1500) identifies the London fort as the departure point, which seems to have initiated a southward relocation of these forces until their subsequent placement by France at St. Laurent.

Second, Boussac and Graville had to gather a large enough force to scare off Talbot who was notorious for impetuous and daring assaults against virtually any size French unit (Talbot dies in a headlong assault against two hundred and fifty cannon laid hub-to-hub at the Battle of Castillon[30]). Anatole France, Alfred Burne, and David Nicolle attempt to solve this sufficiency problem by claiming that Talbot must have seen St. Loup burning before his troops could reach it[31] (the burning actually occurred later that night); therefore, with the fort already lost, Talbot simply turned back to avoid engaging these French troops, which reasoning fails to account for his impetuous nature. Mary Bangs, Wilmot-Buxton, and D'Orliac virtually duplicate this version of France's interpretation of the counterattack by Talbot.

Obviously we can disagree with France's placement of Talbot's forces, as do Gower, Fuller, Smith, Gies, and Castor, who also place them at St Pouair.[32] St. Laurent seems the least likely location for the English to stage their forces at this time. La Hire had attacked St. Pouair just days before, likely to soften it up or to seize it so the French army could more easily

---

[27]Ibid.

[28]Ireland, trans., II, *Journal of the Siege of Orleans*, 67; Parr, I, 116; Lynch, 147; Endore, 97. Of course if church bells were going off, they seemingly would have awakened Joan and d'Aulon who, along with many others, make no mention of them in their accounts. Nevertheless, bells would allegedly signal the arrival of Joan's voices, and she did awaken with her voices talking to her, which may explain why she woke.

[29]Burne, 240.

[30]Ibid., 333-34; Pollard, 1.

[31]France, I, 288-89; Burne, 240; Nicolle, *Orleans, 1429*, 51.

[32]Gower, 53; Fuller, *Decisive Battles*, 484, Smith, 65; Gies, *Joan of Arc*, 75; Castor, 108.

bypass it upon entering the city with their impending northern approach, so there was every reason for Talbot to reinforce and maintain this location. Second, the attack by La Hire against the outpost of St. Loup consisted of Orleans based troops, which suggests that any planning for this raid had taken place prior to the arrival of the French army from Blois and would therefore exclude these units, negating the possibility of Boussac and Graville as either pre-positioned or lying-in-wait. Further, this diversionary tactic against St. Loup had been used before and without the use of any troops from Blois, so there already existed a core of units inside Orleans practiced and proven in successfully attacking St. Loup to allow convoy entry into the city from the south side of the river. These diversionary forces of course excluded those militia who had arrived with Joan from Blois, which forces now served as an unanticipated reserve reaction force of its own.

Finally, in regards to practiced and proven troops, it also makes sense for Boussac and de Rais to command again the security escort for the second supply convoy arriving on the south side of the river. Otherwise, for Boussac to be in charge of the counterattack, he would have had to return to Blois with both Dunois and d'Aulon, which Baucroix's testimony asserts.[33] However, Leon Denis, in 1925, places Boussac on the south side of the river commanding the second supply column arriving from Blois at that time.[34] Parr in 1866[35] and Owen Rutter in 1941[36] also place Boussac and de Rais in command of this second supply column, although Rutter does not state on which side of the river, nor do either use footnotes. Thus, if Boussac led this second convoy, for him to have commanded the French counterattack, he would have had to cross the river, enter Orleans separately and with only a fraction of his escort from the supply convoy, and then immediately ride out to engage Talbot, which seems a stretch.

Clearly, the militia had already gathered or were gathering at the eastern gate when Joan arrived, likely as a part of their expected participation. So if Boussac and Graville were in fact inside Orleans and also organizing their Blois' troops at the Paris gate, we have to accept what amounts to a "general mobilization" with a lot of intermingling and confusion of

---

[33]Murray, 267. Testimony of Simon Baucroix; Parr, I, 104; France, I, 272,281-2. Castor, 107, has de Rais arriving with Dunois from Blois.)

[34]Denis, 62.

[35]Parr, I, 91.

[36]Rutter, 165.

units, or Boussac and Graville would have been too late to intercept Talbot. However, we also know that the French forces were large enough to intimidate Talbot's detachment. For Boussac and Graville to assemble a sufficiently large force on such short notice, after spotting Talbot's movement from St. Pouair toward St. Loup, seems implausible. Most significant to all of this, however, the northern Paris gate, used by Anatole France for this counterattack, had been previously walled up and unavailable except for pedestrian traffic,[37] making it virtually impracticable for dispatching large numbers of troops in a timely manner, especially if mounted.

Thus the French counterattack issuing forth from inside Orleans had to threaten Talbot from the flank or from the rear after the French soldiers departed the city not from the northern Paris gate but from the eastern gate. Therefore Talbot must have been very near St. Loup, allowing La Hire to recognize his danger and permitting de Contes, who was located at the eastern gate, to witness this *preparation to resist.* Any attack from the eastern gate had to include Joan who "advanced against them in all haste."[38]

Finally, because this French tactic had been used before against St. Loup, there is every likelihood of prior planning; therefore, Joan's contingent must have consisted of the volunteer militia which she was consistently allowed to command, although it may have included the persons of Boussac and Graville, and some small gathering of their men, which would then account for the large number of French troops arriving under Joan's banner.

Harriet Parr,[39] Alphonse De Lamartine[40] and Frances Gies[41] are the only authors in agreement with the above portrayal of Joan's attack on St. Loup, although Gies does not name a commander, discuss the composition of forces or explain her reasons for choosing this scenario.

[37]Pernoud and Clin, 226.
[38]Murray, 252. Testimony of de Contes.
[39]Parr, I, 116-17.
[40]Lamartine, 131.
[41]Gies, *Joan of Arc*, 75.

Appendix B

WOMEN IN COMBAT

Joan of Arc was not the first medieval woman to wear full armor or the first to engage in combat, but she was certainly the first "infantrywoman." There are four criteria we can use to make this claim: (1) she was officially commissioned as a soldier by her government; (2) she dismounted her horse and engaged the enemy on foot; (3) she was wounded in combat as proof of her participation; and, (4) her exploits are documented by more than a single source.

Other women of course engaged in war before Joan, but none have met the above criteria. Discounting the Greek legends of the Amazons as beyond proof, most historical examples, by definition, fail to meet even the first criteria: government sanctioning.

Maybe closest to Joan, but outside medieval history, is Lady Fu Hao, a Chinese military commander c.1200 B.C. She was the wife of King Wu Ding of the Shang Dynasty and therefore government sanctioned. She led as many as 13,000 soldiers, but at what age is unknown. Whether she fought on foot or was wounded in action cannot be determined. Her tomb was discovered in 1976. Interestingly, Chinese folklore has another woman, Hua Mulan, disguising herself as a male soldier while fighting in Lady Fu Hao's army for twelve years.

Two Vietnamese women, Trung Trac, and her sister, Trung Nhi, rebelled against Chinese colonists in 40 A.D. Although not sanctioned by any government, at age twenty, Trung Trac, the eldest, successfully led over 10,000 troops, establishing sixty-five strongholds. A large percentage of her leaders were women. The Chinese general, Ma Yuan, with a force of 20,000, defeated her in 42 A.D., and the Trung sisters were beheaded.[1] Again there is no record of any participation in ground combat or any mention of wounds.

The Celtic queen Boudica led the Iceni tribe in a rebellion against the Romans c.60 A.D., and as royalty can be considered government sanctioned. Tacitus claims she led 100,000 combatants who killed 80,000 Romans and Britons. Described only as riding in a chariot, whether she fought on foot or was wounded cannot be determined. The two sources who describe her history, Tacitus and Cassius Dio, are in conflict as to her death, either by suicide after her defeat by the Romans or by illness.

Some might argue that Charles VII did not have a government and that he was nothing more than a landlord, but that is political hyperbole. He maintained all the trappings of a national regime throughout his reign. Political, religious, educational, legal, and economic institutions all remained operational. Thus, this first criterion, a government military commission, validates Joan and virtually eliminates most previous woman warriors.

Even if we discount the first criteria, many women of military note did nothing more than organize and deploy combat forces, much like Yolande in her support of Charles VII. Some women were on the battlefield and even gave commands, but did not personally engage the enemy on foot. Finally, we can find no other women who, after meeting the first two criteria, were subsequently wounded in combat as irrefutable proof of their intimate participation.[2]

Some authors claim that medieval women warriors were well-known and not unusual (for example, Michelet, Haggard and Richey). If so, Christine de Pizan, a contemporary of Joan's, failed to recognize or mention them in her comparisons with Joan. In fact Pizan, as did others, had to reach deep into the Bible to find adequate comparisons. Surely, if so many other women were of sufficient stature, such examples would have leapt to mind without the need for ancient research. Further, none of Joan's contemporaries, male or female, ever mention any other contemporary female warriors who might compare to Joan.

Nevertheless, there are an almost endless number of anecdotes in military history claiming medieval women in combat, which are almost always

---

[1]See Taylor, *The Birth of Vietnam,* 38-40.

[2]One exception might be Hangaku Gozen, a female Japanese warrior samurai, who commanded 3,000 soldiers in a defense against 10,000 attackers. She was wounded by an arrow and defeated.

legendary and Amazonian. A favorite is the story of Brites de Almeida, the baker woman of Aljubarrota, Portugal, who fought in a battle between Castile and Portugal in 1385. Brites was reputed to be a very large woman with six fingers on each hand. After the battle she allegedly ambushed and killed eight Castilian soldiers hiding in her bakery. Maybe true, but likely not. These sorts of tales are everywhere and often designed to feminize and ridicule the losers in battle. Scratching through history to find here or there a female leader or soldier does not equate to a standard of common occurrence as so often professed by those aware of such oddities. With certainty, women were more commonly seen on the battlefield as nurses, runners, suppliers, and scavengers than as combatants, but that does not make them soldiers.

The nearest and most likely comparison to Joan of Arc is probably the fourteenth-century Jean de Montfort who defended her husband's estates, which were not a government entity. The nearest de Montfort came to personally engaging in battle, however, was when she led a cavalry raid against an enemy bivouac, burning down some empty tents. She then fled to a nearby castle where she continued her resistance against besieging forces while sequestered behind stone walls. Her exploits are noteworthy, but not of sufficient notoriety to come to mind as a comparison for Joan of Arc's contemporaries. Further, she never engaged the enemy on foot, nor was she ever wounded as proof of intimate combat.

Another example often cited is the eleventh century Matilda of Tuscany who is said to have been on the field with her troops during their engagements. Her most notable exploit occurred when she led a secret raid on foot from inside her besieged castle at Canossa, Italy, emerging from a hidden entrance concealed in dense fog to surprise and defeat Henry IV's besieging Germans. Although both Jean de Montfort and Matilda may have donned armor,[3] neither was ever in danger of requiring its protections, as neither was ever reported wounded. And Matilda's exploits were apparently of sufficient distance in time and geography not to have entered French minds when Joan of Arc became legend.

If Joan knew of any example after which she might have modeled herself, it would have been Eleanor of Aquitaine who donned male armor to participate in a Holy Crusade with her first husband, Louis VII of France. This crusade was ambushed by Turks at Mount Cadmus, but Eleanor was

[3]Hay, 9, does not believe Matilda ever wore armor.

not a combatant in this action, nor did she ever lead the military force. The crusade failed. Eleanor later married Henry II of England, which of course led to Joan's ultimate need to save France against Eleanor's progeny, which is the most likely reason Joan might have known about her.

Finally, there can be no doubt that some women participated as infantry in some medieval sieges, especially where it concerned the Hussites where they threw down stones on the enemy or otherwise engaged themselves against a military adversary and likely received wounds, but they were never officially sanctioned as a part of the combat force for a bonafide government. Usually they were emergency participants, drafted so to speak, but without any obligation for continued service. The women of Amiens, France, in 1115 A.D., allegedly fought on the frontlines to save their city from attack by a feudal Lord, but lost.[4] Similarly, women were believed to have participated in the attack that ultimately relieved the siege of Compiegne after Joan of Arc was captured,[5] maybe inspired by Joan's example, but again only as emergency participants who left no record of wounds or of sanctioned or continued military service.

Obviously there were medieval women in history who engaged in combat—like the anonymous woman crusader, wrapped in a green cape, who impressed Saladin's men with the accuracy of her longbow before she was finally killed[6]—who nevertheless remain nameless, to which the first sanctioned infantrywoman, Joan of Arc, serves as their best monument and the most accomplished proof of their potential abilities in combat.

[4]Laughlin, 29-30.
[5]Parr, 21; Endore, 215.
[6]Reston, 199.

Appendix C

Joan of Arc's Rules for Miracles

1. **Honor your faith** ~ Commit to your beliefs
2. **Have a vision** ~ Know your purpose
3. **Establish your goals** ~ Be concise
4. **Develop a plan** ~ Be deliberate
5. **Establish standards** ~ Know what you want
6. **Pick your tools** ~ Anticipate your needs
7. **Challenge the status quo** ~ If it ain't working, fix it
8. **Anticipate obstacles** ~ If easy, it would already be done
9. **Choose your fights** ~ Don't waste resources
10. **Be mentally tough** ~ No challenge too great
11. **Be physically fit** ~ The body is part of the brain
12. **Set the example** ~ Be willing to do it yourself
13. **Know the communication system** ~ Listen to the grapevine

14. Market your product ~ Sell, sell, sell

15. Overcome any tests ~ It's OK to win

16. Recognize your duties ~ Be accountable

17. Care for your people ~ Enforce the rules

18. Stay visible ~ Leaders are out front

19. Take responsibility ~ Risk equals reward

20. Have good spurs ~ Stay on the offensive

21. Enter boldly ~ Audacity works

22. Be loyal ~ Trust rewards trust

23. Carry the flag ~ Be the target

24. Accept the risks ~ Courage prevails

25. Ignore the wounds ~ They heal

26. Learn from mistakes ~ The first time

27. Know your enemy ~ Study the problem

28. Be decisive ~ Mean it

29. Persevere ~ Work the problem

30. Stay honorable ~ Trust your conscience

BIBLIOGRAPHY

1. Heresy and Nullification Trials

My primary source for the heresy and nullification trials can be found online or in print:

www.archive.org/stream/jeannedarcmaidof00murruoft

—Murray, T. Douglas (ed.). (1902). *Jeanne D'Arc, Maid of Orleans: Deliverer of France.* New York: McClure, Phillips, & Co.

www. stjoan-center.com

—Murray, T. Douglas (ed.). (1903). *Jeanne D'Arc, Maid of Orleans:*

Deliverer of France. London: William Heinemann.

—*Saint Joan of Arc's Trial of Condemnation.*

—*Saint Joan of Arc's Trial of Nullification.*

Murray, T. Douglas. (ed.). (1907). *Jeanne D'Arc, Maid of Orleans: Deliverer of France.* (New and Revised Edition). London: William Heinemann.

2. Joan of Arc Related Materials

My primary sources for other Joan of Arc related materials came from:

Adams, William Henry Davenport. (1889). *The Maid of Orleans: And the Great War of the English in France.* London: Hutchinson and Co.

Bangs, Mary Rogers. (1910). *Jeanne D'Arc: The Maid of France.* New York and Boston: Houghton Mifflin Company.

Barrett, W.P. (1932). *The Trial of Jeanne D'Arc.* (Coley Porter and Ruth H. Kerr, trans.) New York: Gotham House Inc.

Barstow, Anne Llewellyn. (1986). *Joan of Arc: Heretic, Mystic, Shaman.* New York: The Edwin Mellen Press.

Bartlett, David W. (1859). *The Life of Joan of Arc, the Maid of Orleans*. New York: C.M. Saxton.

Beevers, John. (1981). *St. Joan of Arc (new revised edition)*. Rockford: Tan Books and Publishers, Inc.

Bray, Anna Eliza. (1874). *Joan of Arc and the Times of Charles the Seventh, King of France*. London: Griffith and Farran.

Brooks, Polly Schoyer. (1990). *Beyond the Myth: The Story of Joan of Arc*. Boston: Houghton Mifflin Company.

Brown, Mary Milbank. (1962). *The Secret History of Jeanne D'Arc, Princess, Maid of Orleans*. New York: Vantage Press, Inc.

Buchan, Alice. (1948). *Joan of Arc and the Recovery of France*. New York: The MacMillan Company.

Caddy, Florence. (1886). *Footsteps of Jeanne D'Arc: A Pilgrimage*. London: Hurst and Blackett, Publishers.

Castor, Helen. (2015). *Joan of Arc: A History*. New York: Harper Collins Publishers.

Catherwood, Mary Hartwell. (1897). *The Days of Jeanne d'Arc*. New York: The Century Co.

Cristiani, Leon. (1977). *St. Joan of Arc*. (M. Angeline Bouchard, trans.). Boston: St. Paul Editions.

David-Darnac, Maurice. (1969). *The True Story of The Maid of Orleans*. (Peter de Polnay, trans.) London: W.H. Allen.

Delteil, Joseph. (1926). *Joan of Arc*. (Malcolm Cowley, trans.). New York: Minton, Balch and Company.

Denis, Leon. (1924). *The Mystery of Joan of Arc*. (Arthur Conan Doyle, trans.). London: John Murray, Albemarle Street, W.

DeVries, Kelly. (1999). *Joan of Arc: A Military Leader*. Sutton Publishing.

D'Orliac, Jehanne. (1934). *Joan of Arc and Her Companions*. (Elisabeth Abbott, trans.). London: J.B. Lippincott and Company.

Eliot, George. (1885). *George Eliot's Poetry and Other Studies*. New York: Funk and Wagnalls.

Endore, Guy. (1931). *The Sword of God, Jeanne d'Arc*. New York: Farrar & Rinehart, Inc.

Evans, Joan, trans. *Mystere du Siege d'Orleans.* Oxford: The Clarendon Press.

Evans, Rupert M. (1842). *The Story of Joan of Arc.* New York: D. Appleton & Co.

Fabre, Lucien. (1954). *Joan of Arc.* (Gerard Hopkins, trans.). New York: McGraw-Hill Book Company, Inc.

Ferguson, A.B. Ochiltree. (1923). *The Story of Saint Jeanne d'Arc.* Dublin: The Talbot Press Limited.

For Joan of Arc: An Act of Homage from Nine Members of the French Academy. (1930). New York: MacMillan and Company.

Fraioli, Deborah A. (2000). *Joan of Arc: The Early Debate.* Woodbridge: The Boydell Press.

France, Anatole. (1908). *The Life of Joan of Arc, Vols. I-II.* (Winfred Stephens, trans.). New York: John Lane Company.

Fumet, Stanislas. (1937). *Joan the Saint.* (F.J. Sheed, trans.). New York: Sheed and Ward.

Gies, Francis. (1981). *Joan of Arc: The Legend and the Reality.* New York: Harper and Rowe Publishers.

Gilreath, Shannon D. (2003). *Saint Joan of Arc: History and Reflection.* Baltimore: PublishAmerica.

Goldstone, Nancy Bazelon. (2012) *The Maid and the Queen: The Secret History of Joan of Arc.* New York: Viking Penguin.

Gordon, Mary. (2000). *Joan of Arc.* New York: Viking Penguin.

Gower, Lord Ronald. (1893 - published 2007). *Joan of Arc.* BiblioBazaar.

Gragg, Florence A. (1959). *Memoirs of a Renaissance Pope: The Commentaries of Pius II (abridged).* (Leona c. Gabel, trans. and ed.). New York: G. P. Putnam's Sons.

Grayeff, Felix. (1978). *Joan of Arc: Legends and Truth.* London: Philip Goodall.

Guerin, Andre and Jack Palmer White. (1961). *Operation Shepherdess: The Mystery of Joan of Arc.* London: William Heinemann Ltd.

Guillemin, Henri. (1973). *Joan: Maid of Orleans.* (Harold J. Salemson, trans.). New York: Saturday Review Press.

Haggard, Andrew. (1912). *The France of Joan of Arc.* New York: John Lane Company.

Harrison, Kathryn. (2014). *Joan of Arc: A Life Transfigured.* New York: Doubleday.

Hobbins, Daniel, trans. (2005). *The Trial of Joan of Arc.* Cambridge: Harvard University Press.

Ince, Richard Basil. (1921). *Joan of Arc.* London: William Rider and Son, LTD.

Ireland, William Henry. (1824). *Memoirs of Jeanne D'Arc, surnamed La Pucelle D'Orleans; with the History of Her Times, Vols. I-II.* London: Printed for Robert Triphook.

James, Grace. (1910). *Joan of Arc.* New York: E.P. Hutton and Company.

Jewkes, Wilfred Thomas and Jerome Blanchard Landfield, eds. (1964). *Joan of Arc: Fact, Legend, and Literature.* New York and Burlingame: Harcourt, Brace & World, Inc.

Julleville, L. Petit De. (1901). *Joan of Arc.* (Hester Davenport, trans.). London: Duckworth & Co.

Lamartine, Alphonse De. (1880). *The Standard Series: Joan of Arc.* New York: Funk and Wagnalls, Publishers.

Lamond, John. (1927). *Joan of Arc and England.* London: Rider & Co.

Lang, Andrew. (1909). *The Maid of France.* (new revised edition). New York: Longmans, Green, and Company.

Leary, Francis. (1959). *The Golden Longing.* (Joan of Arc and Rene de Anjou). New York: Charles Scribner's Sons.

Lightbody, Charles Wayland. (1961). *The Judgments of Joan.* Cambridge: Harvard University Press.

Lowell, Francis Cabot. (1896). *Joan of Arc.* Cambridge: The Riverside Press.

Lucie-Smith, Edward. (1976). *Joan of Arc.* New York: W.W. Norton Company Inc.

Lynch, Denis. (1919). *Joan of Arc: The Life-Story of the Maid of Orleans.* New York: Benziger Bros.

Madison, Lucy Foster. (1929). *Joan of Arc: The Warrior Maid.* Philadelphia: The Penn Publishing Company.

Marot, Pierre. (1981). *Joan the Good Lorrainer at Domremy.* (Laure Brissaud, trans.). Colmar: Editions S.A.E.P.

Maxwell, Ronald F. (1994). *Joan of Arc: The Virgin Warrior.* Screenplay: Person to Person Films. C/O Bright Images, Lynchburg, Virginia.

Meltzer, Francoise. (2001). *For Fear of the Fire: Joan of Arc and the Limits of Subjectivity.* Chicago: University of Chicago Press.

Michelet, Jules. (1937). *The Life of Joan of Arc and Other Biographies.* The Spencer Press.

______________ (1967). *Joan of Arc.* (Albert Guerard, trans.). (new revised edition). Ann Arbor: The University of Michigan Press.

Monahan, Michael. (1928). *My Jeanne d'Arc.* New York: The Century Co.

Monstrelet, Enguerrand. (1859). *The Chronicles of Enguerrand de Monstrelet, Vols. I-II.* (Thomas Johnes, trans.). London: Henry G. Bohn.

Nicolle, David. (2001). *Orleans 1429: France Turns the Tide.* London: Osprey Publishing.

O'Hagan, John. (1893) *Joan of Arc.* London: Kegan Paul, Trench, Trubner and Co.

Oliphant, Margaret. (1896). *Jeanne D'Arc: Her Life and Death.* New York: G.P. Putnam's Sons.

Paine, Albert Bigelow. (1925). *Joan of Arc: Maid of France, Vols. I- II.* New York: The MacMillan Company.

___________ (1927). *The Girl in White Armor: The Story of Joan of Arc.* New York: The MacMillan Company.

Parr, Harriet. (1866). *The Life and Death of Jeanne D'Arc: Called the Maid, Vols. I-II.* London: Smith, Elder & Co.

Pernoud, Regine. (1994). *Joan of Arc: By Herself and Her Witnesses.* (Edward Hyams, trans.). (new revised edition). New York: Scarborough House.

___________ (2007). *The Retrial of Joan of Arc.* (J. M. Cohen, trans.) San Francisco: Ignatius Press.

Pernoud, Regine and Marie-Veronique Clin (1998). *Joan of Arc: Her Story.* (Jeremy Duquesnay, trans.). New York: St. Martin's Press.

Pizan, Christine de. (1999). *The Book of Deeds of Arms and of Chivalry.* (Sumner Willard, trans.; Charity Cannon Willard, ed.) University Park: The Pennsylvania State University Press.

Pollard, A.J. (1983). *John Talbot and the War in France, 1427-1453.* New Jersey: Humanities Press Inc.

Purcell, Mary. (1952). *The Halo on the Sword: Joan of Arc.* Maryland: The Newman Press.

Quincey, Thomas De. (1906). *Joan of Arc and the English Mail-Coach.* (Rose Adelaide Witham, ed.). New York: Houghton, Mifflin and Company.

Raknem, Ingvald. (1971). *Joan of Arc in History, Legend and Literature.* Oslo: Universitetsforlaget.

Rankin, Daniel and Claire Quintal, trans. (1964). *The First Biography of Joan of Arc.* (including a translation of part of *The Chronicle of the Cordeliers).* Pittsburgh: University of Pittsburgh Press.

Richards, Laura. (1919). *Joan of Arc.* New York: D. Appleton and Company.

Richey, Stephen W. (2003) *Joan of Arc: The Warrior Saint.* Westport, Connecticut: Praeger Publishing.

Rousseil, Justin. (1925). *St. Joan of Arc: A Study of the Supernatural in Her Life and Mission.* (Rev. Joseph Murphy, S.J., trans.). London: Burns Oates & Washbourne, LTD.

Royden, Agnes Maude. (1918). *Blessed Joan of Arc.* London: Sidgwick and Jackson LTD.

Russell, Preston, (2005). *Lights of Madness: In Search of Joan of Arc.* Savannah: Frederic C. Beil.

Rutter, Owen. (1941). *The Land of St. Joan.* London: Methuen & Co.

Sackville-West, Vita. (1991). *Saint Joan of Arc.* (new revised edition). New York: Doubleday.

Scott, Walter Sidney. (1974). *Jeanne d'Arc.* New York: Harper and Rowe Publishers, Inc.

Searle, William. (1976). *The Saint and the Skeptics: Joan of Arc in the Work of Mark Twain, Anatole France, and Bernard Shaw.* Detroit: Wayne State University Press.

Sermoise, Pierre. (1973). *Joan of Arc and Her Secret Missions.* (Jennifer Taylor, trans.). London: Robert Hale and Company.

Shaw, George Bernard. (1967). *Saint Joan.* (new revised edition). Baltimore, Maryland: Penguin Books.

Shirley, Janet. (1968). *A Parisian Journal 1405-1449.* (translated from the anonymous *Journal d'un Bourgeois de Paris*). Oxford: Clarendon Press.

Smith, John Holland. (1973). *Joan of Arc.* New York: Charles Scribner's Sons.

Southey, Robert. (Undated). *Extracted from The Poetical Works of Robert Southey.* Kessinger Publishing.

Sowers, Pauline B, trans. (1938). *Joan the Maid of Orleans being that portion of the Chronicles of St. Denis which deals with her life and times.* San Francisco: Roy Vernon Sowers.

Spoto, Donald. (2008). *Joan: The Mysterious Life of the Heretic Who Became a Saint.* San Francisco: Harper Collins.

Stanhope, Philip Henry. (1853). *Jeanne D'Arc.* London: John Murray.

Stevens, Charles McClellan. (1918). *Joan of Arc and the Meaning of Her Life for Americans.* New York: Cupples and Leon Company.

Stolpe, Sven. (1956). *The Maid of Orleans.* (Erik Lewenhaupt, trans.). New York: Pantheon Books Inc.

Studer, Paul. (1926). *Saint Joan of Orleans: Scenes from the fifteenth century Mystere du Siege d'Orleans.* (Joan Evans, trans.) Oxford: The Clarendon Press.

Sullivan, Karen. (1999). *The Interrogation of Joan of Arc.* Minneapolis: University of Minnesota Press.

Tallon, Mary Elizabeth, ed. (1997). *Joan of Arc at the University.* Milwaukee: Marquette University Press.

Taylor, Craig, trans. (2006). *Joan of Arc: La Pucelle.* New York: Manchester University Press.

Taylor, Larissa Juliet. (2010). *The Virgin Warrior: The Life and Death of Joan of Arc.* New Haven: Yale University Press.

Trask, Willard, trans. (1996). *Joan of Arc: In Her Own Words.* New York: Turtle Point Press.

Tuckey, Janice. (1880). *Joan of Arc: "The Maid."* London: Marcus Ward and Company.

Twain, Mark. (1989) *Joan of Arc.* (new revised edition). San Francisco: Ignatius Press.

Vale, Malcolm Graham Allan. (1974). *Charles VII.* Berkeley: University of California Press.

Vaughn, Richard. (1966). *John the Fearless.* New York: Barnes and Noble, Inc.

Waldman, Milton. (1935). *Joan of Arc.* Boston: Little, Brown, and Company.

Warner, Marina. (1981). *Joan of Arc: The Image of Female Heroism.* New York: Alfred A. Knopf.

Wheeler, Bonnie and Charles T. Wood, ed. (1996). *Fresh Verdicts on Joan of Arc.* New York: Garland Publishing, Inc.

Whetnall, Arthur. *Joan of Arc.* (19??) London: Robert Culley.

Wilmot-Buxton, Ethel Mary. *Jeanne D'Arc.* (1914). New York: Frederick A. Stokes Company.

Wilson-Smith, Timothy. (2006). *Joan of Arc: Maid, Myth, and History.* London: Sutton Publishing.

Winwar, Frances. (1948). *The Saint and the Devil: Joan of Arc and Gilles de Rais.* New York: Harper and Brothers Publishers.

Wolf, Leonard. (1980). *Bluebeard: The Life and Times of Gilles de Rais.* New York: Clarkson N. Potter, Inc.

Wood, Charles T. (1988). *Joan of Arc and Richard III: Sex, Saints, and Government in the Middle Ages.* New York: Oxford University Press.

Wooster, Nora. (1992). *The Real Joan of Arc?* Sussex, England: The Book Guild Ltd.

Wyndham, Francis M. (1891). *The Maid of Orleans: Her Life and Mission.* London and Orleans.

3. Other Sources

Other source information includes:

Allen, Edgar Leonard. (1970). *From Plato to Nietzsche.* New York: Fawcett World Library.

Allmand, Christopher. (1992). *Henry V.* Los Angeles: California Press.

Antal, John. (2013). *7 Leadership Lessons of the American Revolution: The Founding Fathers, Liberty, and the struggle for Independence.* Philadelphia: Casemate Publishers.

Aries, Philippe and Georges Duby. (1988). *A History of Private Life: Revelations of the Medieval World.* (Arthur Goldhammer, trans.). Cambridge: Harvard University Press.

Army Leadership. (2012). Army Doctrine Reference Publication (ADRP) No. 6-22 (Field Manual 6-22), Headquarters, Department of the Army, Washington, D.C.

Ashe, Geoffrey, Leslie Alsock, C.A. Ralegh Radford, Philip Ratz, and Jill Racy. (1968). *The Quest for Arthur's Britain.* New York: Frederick A. Praeger.

Ashley, Rosco Lewis. (1916). *Medieval Civilization.* New York: The MacMillan Company.

Atkinson, Rick. (2003). *An Army at Dawn: The War in North Africa 1942-1943.* New York: Henry Holt and Company.

__________ (2013). *The Guns at Last Light: The War in Western Europe, 1944-45.* New York: Henry Holt and Company.

Ault, Warren O. (1932). *Europe in the Middle Ages.* New York: D.C. Heath and Company.

Axelrod, Alan. (2001). *Patton on Leadership: Strategic Lessons for Corporate Warfare.* New York: Prentice Hall Press.

Billings, Malcom. (1990). *The Cross and the Crescent: A History of the Crusades.* New York: Sterling Publishing.

Bishop, Morris. (1968). *The Middle Ages.* Boston: Houghton Mifflin Company.

Bisson, T.N. (1991). *The Medieval Crown of Aragon.* Oxford: Clarendon Press.

Bondanella, Peter and Mark Musa, eds. and trans. (1979) *The Portable Machiavelli.* New York: Penguin Books.

Bradley, Omar and Clay Blair. (1983). *A General's Life.* New York: Simon and Schuster.

Bridge, Antony. (1989). *Richard the Lionheart.* New York: M. Evans and Company, Inc.

Buonomano, Dean. (2011). *Brain Bugs: How the Brain's Flaws Shape Our Lives*. New York: W. W. Norton & Company, 2011.

Burne, Alfred H. (1956). *The Agincourt War*. London: Eyre & Spottiswoode.

Cantor, Norman. (1969). *Medieval History (2d edition)*. London: The MacMillan Company.

____________(2002). *In the Wake of the Plague*. New York: Perennial.

Carell, Paul. (1962). *The Foxes of the Desert*. New York: Bantam Books.

Catton, Bruce. (1958). *A Stillness at Appomattox*. New York: Pocket Books, Inc.

Cervantes, Miguel de. (1979). *The Ingenious Gentleman Don Quixote de la Mancha*, (J. M. Cohen, trans.). Franklin Center : Pennsylvania: Franklin Library.

Clausewitz, Carl Von. (1976). *On War*. (Michael Howard and Peter Paret, trans.). Princeton: Princeton University Press.

Cloud, David and Greg Jaffe. (2009). *The Fourth Star*. New York: Crown Publishers.

Connell, Evan S. (1985). *Son of the Morning Star*. New York: Harper and Rowe Publishers.

Collins, Marie and Virginia Davis. (1992). *A Medieval Book of Seasons*. New York: Harper Collins Publishers.

Crane, Stephen. (1964). *The Red Badge of Courage*. New York: Washington Square Press, Inc.

Creasy, Edward S. (1899). *Decisive Battles of the World*. The Colonial Press.

Delbruck, Hans. (1990). *History of the Art of War, Vol. III: Medieval Warfare*. (Walter J. Renfroe, Jr., trans.). Lincoln: University of Nebraska Press.

Dewar, Michael. (1989). *The Art of Deception in Warfare*. Great Britain: David & Charles Publishers.

Du Picq, Ardant. (1946). *Battle Studies: Ancient and Modern Battle*. (Greely, John N. and Robert C. Cotton, trans.). Harrisburg: The Military Service Publishing Company.

Durant, Will. (1939). *The Life Of Greece*. New York; Simon and Schuster.

___________(1944). *Caesar and Christ: A History of Roman Civilization from their beginnings to A.D. 325*. New York: Simon and Schuster.

___________(1950). *The Age of Faith: A History of Medieval Civilization from Constantine To Dante: A.D. 325-1300*. New York: Simon and Schuster.

__________(1953). *The Renaissance: A History of Civilization in Italy from 1304-1576 A.D.* New York: Simon and Schuster.

__________(1954). *Our Oriental Heritage*. New York: Simon and Schuster.

__________(1957). *The Reformation: A History of European Civilization from Wyclif to Calvin: 1300-1564*. New York: Simon and Schuster.

__________(1961.) *The Story of Philosophy: The Lives and Opinions of Great Philosophers of the Western World*. New York: Simon and Schuster.

Durant, Will and Ariel. (1961). *The Age of Reason Begins*. New York: Simon and Schuster.

__________(1968). *The Lessons of History*. New York: Simon and Schuster.

__________(1975). *The Age of Napoleon: A History of European Civilization from 1789 to 1815*. New York: Simon and Schuster.

Earle, Edward Mead., ed. (1973). *Makers of Modern Strategy: Military Thought from Machiavelli to Hitler*. Princeton: Princeton University Press.

Evans, Richard J. (1999). *In Defense of History*. New York: W. W. Norton & Company.

Fremantle, Anne. (1965). *Age of Faith*. New York: Time, Inc.

Fuller, J.F.C. (2001). *Decisive Battles of the Western World*. Great Britain: Cassell & Company.

__________(1936). *Generalship: Its Diseases and their Cure*. Harrisburg, Pa: Military Service Publishing Company.

Gavin, James M. (1985). *On To Berlin*. New York: Bantam Books.

Gazeau, P.F. Rev. (1878). *History of the Middle Ages*. New York: Schwartz, Kirwin & Fauss.

Gies, Francis. (1984). *The Knight in History*. New York: Harper and Rowe Publishers.

Gies, Joseph and Francis. (1979). *Life in a Medieval Castle*. New York: Harper and Rowe Publishers.

__________(1981). *Life in a Medieval City*. New York: Harper and Rowe Publishers.

Gottfried, Robert S. (1983). *The Black Death*. New York: The Free Press.

Grant. R. G. (2013). *Commanders: History's Greatest Military Leaders*. New York: DK Publishing.

Grant, Ulysses S. (1999). *Personal Memoirs*. New York: Penguin Books.

Guderian, Heinz. (1996). *Panzer Leader*. (Constantine Fitzgibbon, trans.) Cambridge: De Capo Press.

Gurevich, A.J. (1985). *Categories of Medieval Culture*. (G.L. Campbell, trans.). Boston: Routledge & Kegan Paul.

Harrison, G.B., ed. (1952.) *Shakespeare: The Complete Works (new revised edition)*. New York: Harcourt, Brace and World, Inc.

Harvey, Robert. (2013). *Maverick Military Leaders: The Extraordinary Battles of Washington, Nelson, Patton, Rommel, and Others*. Delaware: Skyhorse Publishing, Inc.

Hatch, Alden. (1950). *George Patton: General in Spurs*. New York: Modern Literary Edition.

Hay, David J. (2010). *The Military Leadership of Matilda of Canossa*. New York: Manchester University Press.

Heer, Friedrich. (1962). *The Medieval World*. New York: New American Library.

Holmes, George. (1962). *The Later Middle Ages 1272-1485*. New York: W.W. Norton and Company.

__________(1990). *The Oxford Illustrated History of Medieval Europe*. Oxford: Oxford University Press.

House, Adrian. (2001). *Francis of Assisi: A Revolutionary Life*. New Jersey: Hidden Spring.

Howarth, David. (1968). *Waterloo: Day of Battle*. New York: Galahad Books.

__________(1977). *1066: The Year of the Conquest*. New York: Penguin Books.

Jessup, John E. and Robert W. Coakley. (1982). *A Guide to the Study and Use of Military History*. Washington, D.C.: U.S. Government Printing Office.

Jomini, Antoine Henri, de. (1996). *Art Of War.* London: Greenhill Books.

Jones, Archer. (1987). *The Art of War in the Western World.* New York: Oxford University Press.

Kaku, Michio. (2014). *The Future of the Mind.* New York: Doubleday.

Keegan, John. (1984). *The Face of Battle: A Study of Agincourt, Waterloo and the Somme.* New York: Penguin Books.

__________(1987). *The Mask Of Command.* New York: Viking Penguin Inc.

Keen, Maurice. (1984). *Chivalry.* New Haven: Yale University Press.

Kershaw, Alex. (2012). *The Liberator.* New York: Broadway Books.

Koch, H.W. (1978). *Medieval Warfare.* London: Bison Books.

Korda, Michael. (2008). *Ike: An American Hero.* New York: Harper Perennial.

LaCroix, Paul and Walter Clifford Meller. (2002). *The Medieval Warrior.* New York: BCL Press.

Lamb, Harold. (1930). *The Crusades: Iron Men and Saints.* New York: Doubleday, Doran and Company, Inc.

Lieven, Dominic. (2011). *Russia Against Napoleon: The True Story of the Campaigns of War and Peace.* New York: Penguin Books.

Luvaas, Jay. (1966). *Frederick the Great and the Art of War.* New York: The Free Press.

Manchester, William. (1978). *American Caesar: Douglas MacArthur 1880-1964.* Boston: Little, Brown and Company.

___________ (1988). *The Last Lion: Winston Churchill Alone 1932-1940.* Boston: Little, Brown and Company.

Marshall, S.L.A.(1986). *Pork Chop Hill.* New York: The Berkley Publishing Group.

McManus, John C. (1998). *The Deadly Brotherhood: The American Combat Soldier in World War II.* New York: Presidio Press.

Mellenthin, F.W. Von. (1980). *Panzer Battles.* New York: Ballantine Books.

Miskimin, Harry A. (1984). *Money and Power in Fifteenth-Century France.* New Haven: Yale University Press

Murray, Alexander. (1986). *Reason and Society in the Middle Ages*. Oxford: Clarendon Press.

Neillands, Robin. (1990). *The Hundreds Years War*. New York: Rutledge.

Newark, Tim. (1989). *Women Warlords: An Illustrated Military History of Female Warriors*. New York: Sterling Publishing Company, Inc.

Nicolle, David. (1984). *The Age of Charlemagne*. London: Osprey Publishing.

Nietzsche, Friedrich. (1909). *Thoughts Out of Season*. (Adrian Collins, trans.). Edinburgh: T.N. Foulis.

Norman, A.V.D. (1971). *The Medieval Soldier*. New York: Barnes and Noble Books.

Nossov, Konstantin. (2012). *Ancient and Medieval Siege Weapons*. Guilford, Connecticut: Lyons Press.

Nye, Roger H. (1986). *The Challenge of Command*. New Jersey: Avery Publishing Group, Inc.

Oman, C.W.C. (1982). *The Art of War in the Middle Ages. A.D. 378-1515*. London: Cornell University Press.

__________ (1991). *A History of the Art of War in the Middle Ages, Vol. II*. London: Greenhill Books.

Philips, Brigadier General T. R., ed. (1985). *Roots of Strategy*. Harrisburg, Pennsylvania: Stackpole Books.

Powell, General Colin, with Tony Koltz. (2012). *It Worked For Me: In Life and Leadership*. New York: HarperCollins Publishers.

Puryear, Jr., Edgar F. (1984). *19 Stars: A Study in Military Character and Leadership*. Novato, CA: Presidio Press.

RB 101-999: Staff Officers Handbook. (1985). Command and General Staff College, Fort Leavenworth, Kansas.

Reid, Peter. (2008). *A Brief History of Medieval Warfare*. Philadelphia: Running Press Book Publishers.

Reilly, Bernard F. (1993). *The Medieval Spains*. Cambridge: University Press.

Reston, Jr., James. (2002). *Warriors of God: Richard the Lionheart and Saladin in the Third Crusade*. New York: Anchor Books.

Remarque, Erich Maria. (1958). *All Quiet on the Western Front.* Greenwich, Conn.: Fawcett Publications, Inc.

Robertson, William. (1850). *A History of the Middle Ages.* Columbus: William Hall.

Robinson, Major General C.W. (1906). *Wellington's Campaigns—Peninsula to Waterloo 1808-1815.* London: Hugh Rees.

Robinson, John J. (1991). *Dungeon, Fire and Sword: The Knights Templar in the Crusades.* New York: M. Evans and Company, Inc.

Ross, James Bruce and Mary Martin McLaughlin, eds. (1977). *The Portable Medieval Reader.* New York: Penguin Books.

Rothero, Christopher. (1988). *The Armies of Crecy and Poitiers.* London: Osprey Publishing.

Russell, Jeffrey Burton. (1972). *Witchcraft in the Middle Ages.* Ithaca: Cornell University Press.

Ryan, Cornelius. (1966). *The Last Battle.* New York: Popular Library.

Sasso, Claude R. (1982). *Leavenworth Papers No. 6: Soviet Night Operations in World War II.* Fort Leavenworth: Combat Studies Institute.

Sawyer, Ralph D., trans. (2007). *The Seven Military Classics of Ancient China.* New York: Basic Books.

Seward, Desmond. (1978). *The Hundred Years War: The English in France, 1337-1453.* New York: Atheneum.

Sherman, William T. (1889). *Memoirs of William T. Sherman, Vol. I.* New York: D. Appleton and Company.

Soames, Mary, ed. (1998). *Winston and Clementine: The Personal Letters of the Churchills.* New York: Houghton Mifflin Company.

Strayer, Joseph R. (1992). *The Albigensian Crusades.* The University of Michigan Press.

Strayer, Joseph R. and Dana C. Munro. (1959). *The Middle Ages 395-1500.* (new revised edition). New York: Appleton-Century-Crofts, Inc.

Sun Tzu. (1971). *The Art of War.* (Samuel B. Griffith, trans.). (new revised edition). London: Oxford University Press.

Taylor, Keith Weller. (1983). *The Birth of Vietnam.* Berkeley: University of California Press.

Tuchman, Barbara. (1972). *Stillwell and the American Experience in China 1911-45*. New York: Bantam Books.

_____________(1978). *A Distant Mirror: The Calamitous 14th Century*. New York: Alfred A. Knopf.

Tucker, Glenn. (1984). *Chickamauga: Bloody Battle in the West*. Dayton, Ohio: Morningside Bookshop.

Wade, Nicholas. (2009). *The Faith Instinct: How Religion Evolved and Why It Endures*. (New York: Penguin Books), 2009.

Warner, Philip. (1994). *Sieges of the Middle Ages*. New York: Barnes and Nobles Books.

Warner, Rex, trans. (1987). *War Commentaries of Caesar*. New York: New American Library.

Weinschenk, Susan M. (2013). *How to Get People To Do Stuff*. New Riders.

Williamson, Porter B. (1979). *Patton's Principles: A Handbook for Managers Who Mean It*. New York: Simon and Schuster.

Xenophon. (1964). *Anabasis: The March Up Country*. (W.H.D. Rouse, trans.). (new revised edition). Ann Arbor Paperbacks: The University of Michigan Press.

Ziegler, Philip. (1991). *The Black Death*. Wolfeboro Falls: Alan Sutton Publishing.

Index

Made in the USA
Monee, IL
15 January 2026

41612429R00109